THE OCEAN–ENERGY ECONOMY

A MULTIFUNCTIONAL APPROACH

OCTOBER 2023

ASIAN DEVELOPMENT BANK

Contents

Tables

Figures

Boxes

Foreword

This New Ocean–Energy Economy Handbook is one of a series of reference materials on advanced technologies. The goal of this series is to assist the Asian Development Bank (ADB) operations in adopting and deploying advanced technologies in energy projects for its developing member countries, scale up the ADB Clean Energy Program, and move the energy sector closer to achieving its targets in climate finance.

This handbook outlines the types of marine renewable energy sources and how they may be realized into affordable energy for all. The handbook provides an explanation of the technologies and the planning scenarios to make appropriate technology choices in an often-confusing sector.

Case studies will be provided in an attached compendium of technologies provided from industry. The case studies present the project fundamentals including financial, technical, and operational aspects of each deployment.

The emergence of the blue economy has changed the way governments think about our ocean. Developing planning scenarios, increasing awareness through capacity building, and capturing the cross-benefits of related technologies and business approaches are the key themes of this handbook.

The handbook seeks to put some order and common sense to the confusing array of potential solutions that compete for marine spaces. This handbook provides insights into the reasoning behind policy and subsequent technology choices.

May this handbook serve as a helpful reference for ADB operations and its developing member countries as they seek to transition to clean energy and a cleaner environment.

Priyantha Wijayatunga
Senior Director, Energy
Sectors Group

Acknowledgments

This handbook provides an overview of how coastal and island states can adopt a multifunction approach to protect and regenerate their marine estates and exclusive economic zones, while deriving social and economic benefits from them. It was researched under the auspices of the Asian Development Bank's (ADB) Marine Aquaculture, Reefs, Renewable Energy, and Ecotourism for Ecosystem Services (MARES) knowledge and support technical assistance (TA), Promoting Sustainable Energy for Developing Member Countries for Asia and the Pacific (TA 6619-REG). The TA project will facilitate an integrated development approach combining the four areas of MARES and emphasizing value-added economic activities.

The MARES concept, handbook, and associated publications are the result of the team effort led by Stephen Peters, senior energy specialist (Waste-to-Energy), Energy Sector Office, Sectors Group (SG-ENE), with inspiration and support from Dan Millison, consultant, SG-ENE. We are also thankful to the following ADB staff for their valuable technical and operational inputs: Kee-Yung Nam, principal energy economist, SG-ENE; Rafayil Abbasov, senior specialist, SG-ENE; Len George, principal energy specialist, SG-ENE; Gary Krishnan, senior country specialist, Regional Cooperation and Operations Coordination Division, Southeast Asia Department; Maria Dona Aliboso, operations analyst, SG-ENE; Ellen Paul, senior country officer, Pacific Department (PARD); and Francesco Ricciardi, senior environment specialist, Office of Safeguards.

We extend our thanks to the many experts who contributed to researching and drafting the text, notably the MARES Expert Group, particularly Michael Abundo, chief executive officer (CEO), OceanPixel Pte Ltd.; Tom Bowling, CEO, Biota Inc.; Scott Countryman, executive director of the Coral Triangle Conservancy; Nguyen Dinh, head of Hydrogen and Principal in Ireland, Offshore Wind Consultants; Jack Dyer, Blue Economy Future SA; Andy Hamflett, co-founder and director of NLA International Ltd.; Gary Hesling, associate director, NLA International Ltd.; Gregor Hodgson, independent consultant and peer reviewer; Alex Rogers, director of Science, REV Ocean; Sabina Rustamova, social safeguards specialist; and Ivory Vogt, program manager, Climate and Resilience, Sustainable Travel International.

Our in-country MARES representatives in the Republic of the Marshall Islands and the Republic of Palau provided critical local information, contacts, and guidance. They are Don Hess and Setoki Qalubau, both from the Marshall Islands, and Fabian Iyar from Palau.

We are also very grateful to the many organizations, companies, and people who contributed as speakers and panelists to the MARES webinar series and other events such as the ADB Asia Clean Energy Forums in 2021 and 2022. They include Ferhat Acuner, general manager, Navtek Naval Technologies Inc.; Richard Argall, technical director, Makai Ocean Engineering, Inc.; Karina Barquet, senior research fellow, Stockholm Environment Institute; Will Bateman, CEO, CCell Renewable UK; Alix Burrell, principal investment specialist, Infrastructure Finance Division 2, ADB; Christine P. Chan, senior advisor to the Vice-President, ADB; Tom Chi, founding partner, At One Ventures; Loke Ming Chou, National University of Singapore; Cindy Cisneros-Tiangco, principal energy specialist, SG-ENE, ADB; Agostinho Miguel Garcia, ADB consultant, Preparing Floating Solar Plus Projects under the Pacific Renewable Energy Investment Facility (TA 6680-REG); Emily Hazelwood and Amber Sparks of Blue Latitudes; Yasuyuki Ikegami, director of the Institute of Ocean Energy, Saga University, Japan; Kirsten Isensee, programme specialist, Ocean Carbon, Intergovernmental Oceanographic Commission of United Nations Educational, Scientific and Cultural Organization; Stefan Kraan, chief scientific officer of The Seaweed Company, Ltd.; Marcel Kroese, marine biologist consultant, International Non-Government Organizations; Benjamin Martin, project manager, Xenesys Inc.; R. Duncan McIntosh, senior regional maritime specialist, ADB; President James Michel, former president of the Republic of Seychelles; Jan Newton, senior principal oceanographer and affiliate professor Oceanography, University of Washington; Santosh Kumar Srirangam, HSL Constructor Pte Ltd.; Pierre Rousseau, independent consultant, Sustainability and Finance; Gürdoğar Sarıgül, senior consultant, Environment, Climate Change and Maritime Decarbonisation; Leo Ban Tat, ECO-ARK, Aquaculture Centre of Excellence Pte Ltd.; Harry Thomson, future energy project manager, Shetland Islands Council; Monica Verbeek, executive director, Seas at Risk; and Steve Widdicombe, director of science and deputy chief executive, Plymouth Marine Laboratory.

We are grateful to ADB team members Fely Arriola and Imee Aquino, both consultants, who provided guidance and support throughout.

Special thanks also to Nick Lambert, co-founder and director of NLA International, for facilitating the coordination among these different companies and organizations. Nick's extensive knowledge of the blue economy was underpinned by his carrier in the British Royal Navy, leaving his post as chief hydrographer as a rear admiral. Nick's leadership was a key success factor for the MARES TA project. Thanks also go Nick's other directors at NLAI, Jonathan and Andy Hamflett who stepped up to support the business plan competition and workshops.

Thanks go to the staff at the Centre for Indonesia-Malaysia-Thailand Growth Triangle (CIMT-GT) who partnered with ADB to host the MARES High-Level Investor Forum on 7 February 2023. Special thanks goes to Executive Director Pak Firdaus Dahlan. Thanks also to the CIMT Business Council for their support, especially Amzar Azhar.

Thanks should also be extended to the team who managed the DevAsia dataroom with over 25 webinars by loading these and further technical reports and other content onto the ADB Data Room created for the MARES program. Thanks to Maria April Dela Cruz for her efforts with Kris Guico in editing and uploading webinars.

This handbook presents the nontechnical reader the best available marine renewable energy technologies; provides explanation of the science and engineering involved; and identifies the policy gaps and the commercial outcomes leading to social and environmental needs. It will enable developing member countries to make decisions on the optimal marine renewable energy technology that is most applicable and suitable to their climate and environment, aligned to national policy and strategy.

We hope that through this handbook, readers would be better able to comprehend the current state of development and how to move toward a regenerative economy while addressing the energy production challenges facing our civilization.

Abbreviations

ADB	Asian Development Bank
AI	artificial intelligence
CAGR	compound annual growth rate
COTS	crown-of-thorns starfish
COVID-19	coronavirus disease
DMC	developing member country
EEZ	exclusive economic zone
EU	European Union
FAO	Food and Agriculture Organization of the United Nations
FPV	floating solar photovoltaic
GDP	gross domestic product
HATT	horizontal axis tidal turbine
LCOE	levelized cost of energy
LCOH	levelized cost of hydrogen
MARES	Marine Aquaculture, Reefs, Renewable Energy, and Ecotourism for Ecosystem Services
MRE	marine renewable energy
NGO	nongovernment organization
ORE	ocean renewable energy
OTEC	ocean thermal energy conversion
PIC	Pacific island country
PRC	People's Republic of China
PRO	pressure-retarded osmosis
PV	photovoltaic
RED	reverse electrodialysis
SDG	Sustainable Development Goal
SIDS	small island developing state
SWAC	seawater air-conditioning
TRL	technology readiness level
UK	United Kingdom
UN	United Nations
US	United States

Weights and Measures

GW	gigawatt
GWh	gigawatt-hour
km	kilometer
km^2	square kilometer
kW	kilowatt
kWh	kilowatt-hour
m	meter
m/s	meter per second
m^2	square meter
m^3	cubic meter
MW	megawatt
MWh	megawatt-hour
TW	terawatt
TWh	terawatt-hour

Introduction to the Handbook

The Asian Development Bank (ADB) seeks to support blue economy opportunities by investing in "regenerative" activities that are economically sound and benefit the health of our ocean.

The term "regenerative" is used to mean "activities which restore and/or enhance ecosystem services." This document presents an overview of, and interim findings from ADB's Marine Aquaculture, Reefs, Renewable Energy, and Ecotourism for Ecosystem Services (MARES) project. MARES is exploring how coastal and small island developing states can adopt a novel, multiple-use approach to better manage their coastal and marine areas and exclusive economic zones.

ADB supports its developing member countries in harnessing renewable energy from the ocean to diversify and boost mariculture, promote marine ecotourism, and rehabilitate coral reefs as a source of food and coastal protection.

This handbook attempts to present information on a wide range of blue economy domains and areas of practice, at the same time as setting out existing, emerging, and aspirational linkages between them. It is therefore structured in the following way to guide the reader through this complexity.

Section 1 (Seascape) introduces the marine and maritime space. This section starts out by examining some of the challenges facing our seas and oceans (and the sectors, people, and ecosystems that depend on them). Chapter 2 then highlights how various maritime sectors and areas of marine management are now being seen more holistically through a blue economy lens. Chapter 3 provides background to the key sectors identified for analysis within the MARES approach (marine aquaculture, cultivated reefs, and ecotourism), highlighting the investments and innovative approaches of greatest interest within each.

Section 2 (Powering the Future) introduces the key concepts related to marine renewable energy (MRE) and explores how they might enable multifunction marine and maritime projects. Chapter 4 explains how MRE is central to the crosscutting delivery of several blue economy benefits, before providing short introductions to a range of MRE technologies, from offshore wind to tidal power. Chapter 5 delves a little deeper by introducing more recent MRE innovations as well as exploring some of the potential risks to be managed when introducing the various types of technology. Finally, Chapter 6 explains how green hydrogen has the potential to be an important catalyst for coastal states by allowing them to convert their marine energy to electricity, and then to export that to an entirely new market. The chapter ends with a number of case studies of green hydrogen in action. Chapter 7 begins to weave these disparate strands into the MARES approach, highlighting how multifunction marine and maritime projects have the potential to be greater than the sum of their parts. Examples are provided of multifunction MRE approaches, from hybrid floating wind and wave platforms to the integration of renewable energy sources into aquaculture operations.

Section 3 (Ocean Nations in Control) moves from the global to the local scenario by examining the challenges and opportunities that present themselves when attempting to conceptualize the development of MARES projects within specific island settings. Chapter 8 presents a detailed examination of specific development possibilities in the Marshall Islands. It provides the backdrop to MRE possibilities within the Pacific state; articulates some key blue economy sector opportunities; identifies and analyzes specific geographical sites that may be most amenable to the MARES approach; and suggests the kind of multifunction projects that may have the greatest impact, considering all the contributing factors above. Chapter 9 repeats this process for Palau and supports further exploration there. Exploration of this additional state helps to clarify crosscutting themes and differentials when considering the MARES projects. Finally, Chapter 10 takes that narrative up a level by revealing the process that MARES project specialists developed to complete the site-specific analysis undertaken in Chapters 8 and 9. This spells out the principles and criteria developed by the MARES team to assess candidate technologies and pilot sites, aligned to the range of identified risks to be considered and managed. In essence, this provides a new checklist for coastal and island states, technology developers, and private or institutional investors, to initiate the exploration of the MARES multifunction projects themselves.

The final section draws conclusions from the broad range of analysis undertaken, and points to next steps.

The summary information presented in this handbook is underpinned by more detailed research and analysis, especially related to scoping potential projects in the Marshall Islands and Palau. Highlights from the National Indicative MARES Plans are presented in Chapters 8 and 9.

Over a month-long process, nine projects were identified to showcase the MARES approach. These projects were mentored and reviewed by experts for project showcase at a high-level investor forum in February 2023. An extract of the proceedings of this forum and the project showcase information are available in Appendix 1, with insights into how to operationalize the MARES approach.

Executive Summary

The ocean is an economic powerhouse supporting 90% of global trade, enabling an abundance of maritime industries, and providing sustenance for millions of people worldwide. Moreover, the true extent of marine carbon capture capabilities is only now beginning to be appreciated and further prioritized, as the climate change urgency becomes more acute.

However, our marine spaces and the ecosystems, economies, and livelihoods that depend on them are undergoing rapidly increasing pressure. Stressors include climate change and increased water temperatures; more powerful hurricanes or typhoons featuring higher wind speeds, storm surges, and wave heights; ocean acidification; marine pollution; sedimentation; overfishing, destructive fishing, illegal, unreported, and unregulated fisheries; poorly planned coastal development; increased population growth and demand for all coastal and marine resources; and sea level rise.

Many of these issues are already causing human suffering and economic losses, leading to the collapse of marine ecosystems covering thousands of square kilometers. Without intervention, commercial fishing stocks will continue to deplete, significantly exacerbating food security. Over half a billion people will be threatened by sea level rise by the end of the century. While these challenges patently need to be addressed with urgency, the United Nations (UN) Sustainable Development Goal (SDG) most aligned to these challenges (SDG 14 – Life Below Water) is widely acknowledged to be one of the least prioritized themes.

A Positive Outlook

At first look, this presents a bleak picture for islands and coastal states. However, many are not yielding, instead offering vibrant leadership, engaging globally to subvert a narrative rooted in inevitable decline, and articulating how their marine resources can provide a platform for a more positive, climate-resilient future for all. This positive approach encourages all stakeholders to view the oceans as an opportunity, rather than just a crisis to be mitigated. It also aligns well with the continued emergence of the blue economy, which looks to promote ocean policies that foster economic growth while at the same time regenerating ocean health in a manner that is consistent with social equity and inclusion.

A truly holistic blue economy approach will enable governments to deliver long-term socioeconomic benefits through sensitive marine stewardship, which will include targeted investment in regenerative ocean resources.

The increasing confidence in the blue economy is emerging for a number of reasons. First, there is growing understanding of the breadth of positive outcomes a well-managed marine area can support—not just regional and national targets related to jobs, economic growth, and environmental protection, but aligned to broader agenda such as the SDGs; the UN Environment Programme Finance Initiative; the Sustainable Blue Economy Finance Principles developed by the European Commission and others; and nationally determined contributions for climate change action under the Paris Agreement and the UN's Framework Convention on Climate Change.

Such confidence is also encouraging greater innovation in several discrete ocean sectors including mariculture, cultivated reefs, nature-based coastal defenses, and marine ecotourism, each of which offers new opportunities for island and coastal states. In addition to innovation within individual sectors, several new macro approaches to the blue economy are bedding in, including investment in blue innovation clusters, the development of policies to address climate change, the rise of ocean entrepreneurship, sustainable impact investment to finance blue ventures, and governments adopting a range of new initiatives to create enabling environments and develop the skill sets necessary for blue innovation.

Given this growing surge in impact investment and the potential to upscale innovations, though, the scale of ambition does not yet match the size of the opportunity for both small island states and nations with large exclusive economic zones.

The challenge remains that unless proper strategic coordination is achieved, the true potential of the blue economy will not be realized. Coordinated action is repeatedly highlighted as being of critical importance and needs to be embedded at all levels to maximize the benefits of international strategic leadership, the value of the blue economy approach, and to ensure that individual sector initiatives are complementary.

Marine Renewable Energy: The Heart of the Blue Economy

The Marine Aquaculture, Reefs, Renewable Energy, and Ecotourism for Ecosystem Services (MARES) project has been initiated to illustrate how coordinated efforts can have greater tangible impact within ocean domains. Its outward-facing focal point is to find and support multifunction marine and maritime projects that feature the integration of marine renewable energy (MRE) platforms, as well as to identify ways to enhance the enabling conditions to support such innovations to scale.

To maximize the potential of such initiatives, the MARES project also looks to highlight two specific strategic opportunities that when seen together provide greater confidence in this approach.

The first blue economy strategic opportunity that underpins the MARES approach is the immediate and growing potential to accelerate the harnessing of MRE. Theoretically, estimates suggest that the global annual potential of wave energy could meet the entire global energy demand. Tidal energy alone has the potential to power over 20 million American homes.

While it is true that this potential is a long way off being realized, two trends suggest that this is beginning to change. First, there is growing visible recognition of the importance of including MRE within blue economy road maps and ocean strategies to realize multiple benefits. Aside from the obvious advantages of more cleaner energy sources, crosscutting gains are identified as (i) being able to power systems that observe, monitor, and protect the oceans; (ii) the cost-effective provision of more clean water; and (iii) the development of many more businesses across the MRE supply chain, in turn leading to more sustainable livelihoods. Crucially, the emerging golden thread weaved into this thinking is that well-designed and effective MRE planning can support a wide range of industries, including ecotourism, sustainable aquaculture, and seawater mineral extraction.

That broader strategic understanding alone creates a stronger platform for greater investment in MRE development. It is matched by promising technological progress in all the main MRE technology segments. Offshore wind technologies are developing at pace (especially as floating wind platforms push further out to sea), as is the broader supply chain where artificial intelligence and advanced robotics, in particular, are being deployed successfully to reduce operational and maintenance costs. Individual floating solar platforms are beginning to attract billions of dollars of investment. Innovations and design rationalization in wave energy conversion and ocean thermal energy conversion systems are propelling them up the technology readiness level scale.

Green Hydrogen: Catalyzing the Marine Renewable Energy Opportunity for Smaller States

MRE developments offer greatest promise to larger, more developed nations that have the capacity demands to support investment decisions, where sound economic returns can be visualized. The MRE picture is more complex within smaller coastal and island states. These nations potentially have the most to offer (in terms of marine estate and/or exclusive economic zone potential prospects to be made available for renewable energy and multifunction marine projects) and the most to gain in creating new socioeconomic opportunities to counter the shock of the coronavirus disease (COVID-19); to create revenues to invest in more regenerative, just transition activities; and to reduce climate change impacts that threaten their very existence in the short to medium term.

However, as they are often typified by smaller populations that do not require the kind of energy at scale that will unlock commercial investments, the opportunity for MRE development flounders.

This status quo could be transformed with the emergence of marine green hydrogen. **This second major strategic opportunity potentially creates an encouraging new blue horizon, by allowing large ocean states to utilize their sea spaces not only to meet local energy demands, but also to convert energy to electricity, and then to export that to an entirely new market (a "power-to-x" capability)**. Both policy and commercial drivers are demanding alternative marine fuels, for example, which creates significant new market possibilities in this realm.

Marine green hydrogen, a new *portable* energy storage solution, could allow ocean states to utilize their significant coastal and marine areas not only to meet local energy demands, but also to convert excess MRE to exportable hydrogen. This promising market model, which did not exist before, could unlock the financing required to speed up adoption of these technologies and secure a key regenerative cornerstone of many nations' blue economies.

This is no longer theoretical. Around the world, examples are emerging of the effective use of green hydrogen—enabling energy sustainability in small coastal communities, helping to reduce carbon emissions in port operations through the commercialization of hydrogen-powered barges, and developing plans to export marine green hydrogen once capacity outstrips immediate local demand.

Cooperation in Action: The MARES Approach

These broader signs of encouragement—new investment flowing into the blue economy and MRE output progressing in line with the investment narrative changing around the new green hydrogen opportunity—provide the backdrop for more tactical endeavors.

The immediate opportunity is to develop a much broader set of multifunction projects that make the most effective use of available sea space, providing the greatest positive effect on local communities (including ensuring that any resulting prosperity is equitably distributed) and that result in the lightest environmental impact.

This is the MARES approach: MRE projects thrive under a coherent strategic policy framework, providing energy for local use and for powering multifunction projects, and, at the same time, serving as a new energy export market.

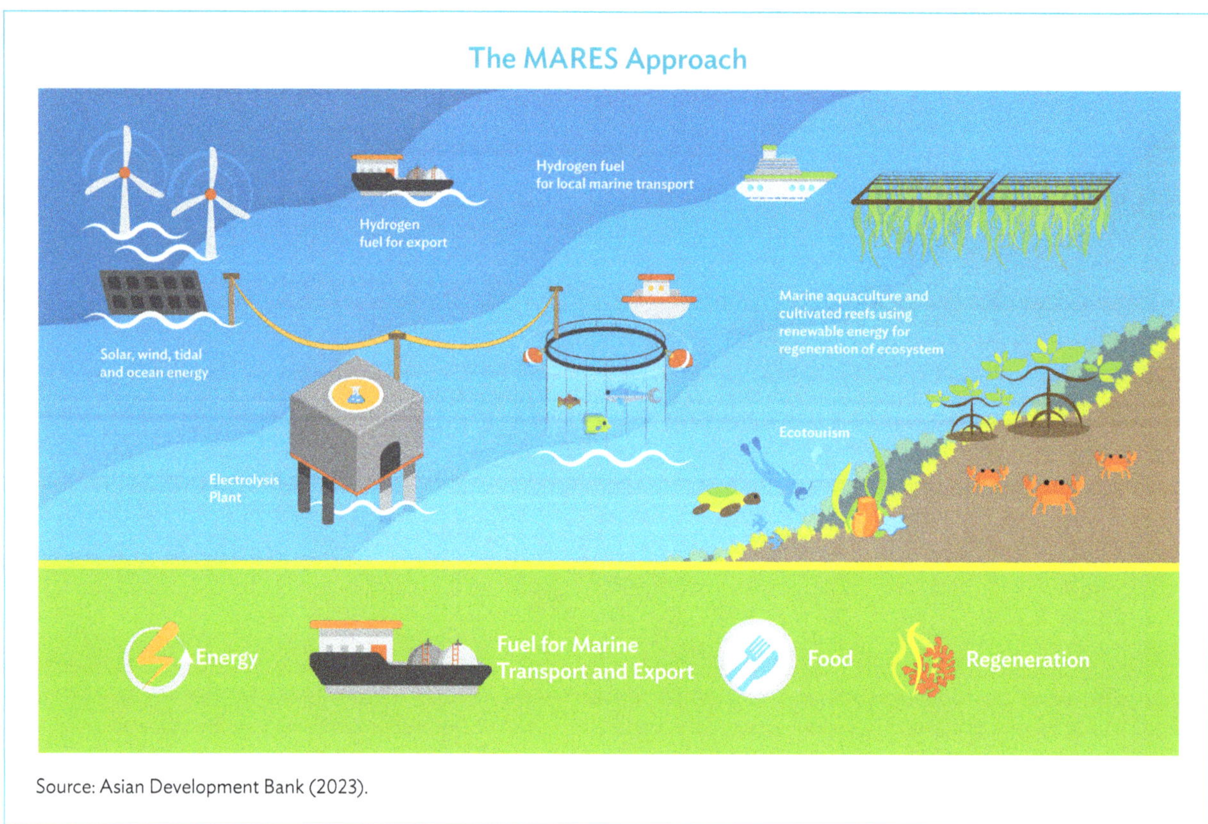

Source: Asian Development Bank (2023).

Many exciting innovations are being developed in marine aquaculture, cultivated reefs, and nature-based marine tourism. The new market opportunities described here create the right conditions for more ambitious scaling—and progress that fully integrates the many benefits of MRE.

More resources, effort, and thinking are required now, not just to see how marine activities can be planned to live comfortably side by side, but how they can properly integrate with each other, how they can actively support each other, and how they can become more than the sum of their parts—thus adopting a true blue economy approach as opposed to a rebadging of existing marine and maritime sectors.

An Island Focus

There are immediate opportunities to take this approach to island states, both to provide near-term benefits and to provide case studies for others to utilize. Analysis of potential in the Marshall Islands and Palau has led to the development of a coherent and replicable set of principles and decision-making tools to ensure that putative multifunction marine and maritime projects are apprised of the existing policies in place—sensitive to local environmental conditions so that the right sites can be selected and in tune with social drivers and community needs. Sensibly considering these opportunities in the round will provide the strongest starting point for MARES projects. In the next phase of operation, the program will look to build on these insights and identify specific and tangible steps to put them into action.

The business plan competition results showed that there are already pioneer projects using the MARES approach now with technology that is commercially available.

1

SEASCAPE:
AN INTRODUCTION
TO THE BLUE ECONOMY

CHAPTER 1

Oceans in Crisis

This introductory chapter provides a brief background on the state of ocean industries and marine ecosystems.

Introduction

Our seas and oceans are an incredibly vital part of our lives. They cover 72% of the earth's surface, support over 90% of global trade, and constitute more than 95% of the biosphere.[1] Ocean assets that provide employment, revenue, and climate change solutions include the following:

- **Capture fisheries and marine aquaculture.** The State of World Fisheries and Aquaculture report of the Food and Agriculture Organization of the United Nations (FAO) estimated 2020 production of around 178 million tonnes. It was estimated that the total first sale value of the global production was $406 billion, of which $141 billion was for capture fisheries and $265 billion for aquaculture.[2]
- **Shipping and ports.** By the end of 2020, around 11 billion tons of goods were transported by ship, representing around 1.5 tons per person based on the current population.[3]
- **Shipbuilding.** In 2018, the global shipbuilding market amounted to roughly $114.3 billion.[4]
- **Marine and coastal tourism.** The tourism sector has been temporarily slowed by the coronavirus disease (COVID-19) but is expected to be booming again soon; there are 3.2 million jobs in the maritime tourism sector in Europe alone.[5] Of greatest interest to the blue economy, around 80% of all global tourism takes place in coastal areas, and among the most popular destinations are beaches and coral.[6]
- **Offshore oil and gas.** This is a multitrillion-dollar titan of the ocean economy, which combines drilling, installation, decommissioning, support vessels, pipelines, process services, floating production storage, submarines, remotely operated vehicles, subsea-well access, seabed survey, environmental impact assessment, mooring systems, and subsea pumps.

Other significant emerging industries include safety and surveillance; marine biotechnology; high-tech marine services; communications, research, education maritime defense and security; marine and maritime research and education; and marine seabed mining, hydrogen, and/or liquefied natural gas —alternative marine fuels and decarbonization, desalination, blue carbon, and ecosystem restoration.

---

[1]   C. De Beukelaer. 2020. Ships moved more than 11 billion tonnes of our stuff around the globe last year, and it's killing the climate. This week is a chance to change. *The Conversation.* 16 November.
[2]   FAO. 2020. *The State of World Fisheries and Aquaculture.* Rome.
[3]   International Chamber of Shipping. Shipping and World Trade: Driving Prosperity.
[4]   European Commission. Shipbuilding Sector.
[5]   European Commission. Coastal and Marine Tourism.
[6]   United Nations World Tourism Organisation (UNWTO). 2014. *UNWTO Tourism Highlights.* 2013 Edition.

Our oceans play a significant role in every aspect of life on earth. Most notably, they can regulate climate change by acting as a giant carbon sink. Indeed, since the Industrial Revolution, the ocean has absorbed about a third of the carbon dioxide generated by human activities.

Shocks to the System

Our oceans are used directly and indirectly by billions of people, and they are vital to our global economy and trade. Unfortunately, human impacts on the oceans have increased with population growth and our increasing commercial use of their resources from fish to cooling water for power plants. The world's seas and oceans and the ecosystems, economies, and livelihoods that depend on them are under great stress from anthropogenic impacts especially climate change, ocean acidification, and sea level rise leading to the collapse of marine ecosystems.

Climate change has caused increased water temperatures; more powerful hurricanes and typhoons featuring higher wind speeds, storm surges, and waves; ocean acidification; marine pollution; sedimentation; overfishing, destructive fishing, illegal, unreported, and unregulated fisheries; poorly planned coastal development; increased demand for all coastal and marine resources; and sea level rise. All these issues are already leading to the collapse of marine ecosystems covering thousands of square kilometers (km^2), such as the Great Barrier Reef, and need to be urgently addressed.[7]

Around the globe, there are approximately 39 million fishers, and the great majority of them are small-scale operators (footnote 2). Their numbers outweigh those in the industrial fisheries,[8] oil and gas, shipping, and tourism sectors combined.[9] If post-harvest activities are included, about 50% of workers are estimated to be women (footnote 2). Many of these people remain among the poorest people in society and their livelihoods are threatened by overextraction and climate change.[10]

Productive wild-caught fisheries depend on a healthy ocean.[11] However, more than a third of major commercial fish species are being fished at biologically unsustainable levels,[12] and an additional one-third are fished at their maximum sustainable level. With approximately 3.2 billion people around the world currently relying on fish for nearly 20% of their animal protein, it is difficult to accommodate a globally increasing population that is reliant on wild-caught fish protein. If trends continue, it is estimated that there will be almost no stocks left for commercial fishing by 2048 in Asia and the Pacific region.

There are more serious problems than fisheries, however. Sea level rise, caused by global warming and melting ice sheets, is already regularly causing coastal erosion and inundation in numerous countries from the United States (US) to Bangladesh. By 2100, over 600 million people will be threatened by flooding from sea level rise.[13]

[7]   United Nations Environment Programme (UNEP). 2012. *Green Economy in a Blue World*. Nairobi; The Economist Intelligence Unit. 2015. *The Blue Economy: Growth, Opportunity and a Sustainable Ocean Economy*. Briefing Paper for the World Ocean Summit 2015. Cascais, Portugal. 3–5 June; UNEP. 2016. *Blue Economy Concept Paper*. Nairobi; and World Bank. 2017. *The Potential of the Blue Economy: Increasing Long-term Benefits of the Sustainable Use of Marine Resources for Small Island Developing States and Coastal Least Developed Countries*. Washington, DC.

[8]   L. C. L. Teh and U. R. Sumaila. 2011. Contribution of Marine Fisheries to Worldwide Employment. *Fish and Fisheries*. 14 (1).

[9]   H. Smith and X. Basurto. 2019. Defining Small-Scale Fisheries and Examining the Role of Science in Shaping Perceptions of Who and What Counts: A Systematic Review. *Frontiers in Marine Science*. 6. 236.

[10]   V. W. Y. Lam et al. 2020. Climate Change, Tropical Fisheries and Prospects for Sustainable Development. *Nature Reviews Earth & Environment*. 1 (9). pp. 440–454.

[11]   UNEP. 2012. *Green Economy in a Blue World*. Nairobi.

[12]   FAO. 2021. *The State of World Fisheries and Aquaculture*. Rome.

[13]   E. Kirezci et al. 2020. Projections of Global-Scale Extreme Sea Levels and Resulting Episodic Coastal Flooding over the 21st Century. *Scientific Reports*. 10 (1). 11629.

In the 200 years since the Industrial Revolution began, the manufactured increase of carbon dioxide in the oceans represents a 30% increase in acidity—an acidification rate that is about 10 times faster over the last 55 million years. Higher acidity is already having a significant impact on marine life. In some locations, young oysters have been unable to create a shell. Eventually, any ocean animal with a bone skeleton or shell will be threatened.

Over the past half century, coastal and marine tourism has expanded in popularity. The Organisation for Economic Co-operation and Development (OECD) predicts that coastal and marine tourism will be the largest sector of the global ocean-based economy by 2030, generating a total revenue of $777 billion globally and employing around 8.6 million people.[14] Tourism provides an important source of livelihood for many island nations and coastal economies. Coastal and marine tourism depend on clear blue water for snorkeling, scuba diving, and swimming, and tourists prefer white sand beaches. But as more people live near the coast or come as tourists, the human impacts increase. Sedimentation and marine pollution have negatively affected many previously pristine areas from Guam to Boracay Island, Philippines. Tourism authorities are aware of these problems and have taken an active role in working with companies to find and implement solutions using certification programs such as Blue Flag. Where implemented, these solutions have helped reduce such problems at the local scale and brought attention to them on a larger geographic scale.

The negative impacts of global climate change caused by human use of fossil fuels are now in the news daily—forests are burning from Australia to France to California, and coral reefs bleaching and dying around the tropics. In 1997, Reef Check carried out the first global survey of coral reefs and reported to the 2002 World Summit on Sustainable Development that there was a global coral reef crisis.[15] This was so surprising at the time that few scientists believed it, and the public perhaps wondered how this might affect them. Now, after losing much of the live coral on the Great Barrier Reef of Australia as well as from coral reefs throughout the world, it is widely understood that global warming is heating up the oceans so much and so fast that it is often simply too hot for corals to survive. About half of the world's coral reefs and mangrove forests that normally help protect the coast have disappeared over the past 50 years. By 2052, 90% of coral reefs may disappear, and oceans might have more plastic than fish in terms of weight.[16]

The Funding Gap

Given the high value of the ocean economy, the dependence of billions of people on it, and the threats to ocean health and biodiversity, clearly more investment in regenerative commercial ocean ventures is needed. There is a large funding gap. KPMG's Survey of Sustainability Reporting identified the United Nations (UN) Sustainable Development Goal (SDG) 14 (Life Below Water) as one of the least prioritized SDGs, with around 18% of companies marking it as a focal priority, despite its estimated funding requirement of $174.52 billion per year.[17] This major financial shortcoming presents a barrier to all stakeholders—not just coastal and island states, emerging economies, and the millions who rely on the ocean for livelihoods, but the entire global ocean economy. Poor information and a lack of consensus over which innovations to support, replicate, and scale up, combined with a lack of standardized methodology, inconsistent impact indicators, and comparatively few investments and/or market-ready solutions among industries, are creating a fundamental financing obstacle that needs to be overcome.[18]

14   OECD. 2016. *The Ocean Economy in 2030*. Paris. OECD Publishing.
15   ReefCheck Worldwide. Coral Reef Program.
16   World Economic Forum. 2016. *The New Plastics Economy: Rethinking the Future of Plastics*. Geneva.
17   KPMG. 2020. *The Time Has Come: The KPMG Survey of Sustainability Reporting 2020.*
18   A. Cowell. 2021. *The Blue Economy – A Drop in the Ocean*. KPMG Blog. 3 June.

Traditionally, funding to solve many environmental problems has been provided in the form of grants from private foundations. The finance community has viewed environmental issues as a compliance problem requiring the prevention of negative environmental impacts and costs. Thinking of externalities, instead of opportunities, was dominant. Private sector funding has evolved such that impact investments with social and environmental goals are part of the desired results of the investments, in addition to profits, albeit at a reduced return and with a longer tenure.

It is possible to achieve synergies among a combined group of projects from different sectors, such as marine renewable energy (MRE), mariculture, and coastal protection, with a symbiotic approach. This requires expertise and capacity to demonstrate and document design benefits, understand funding options, and execute projects. Any project team needs to be conscious and experienced in related risks, understand innovations, exploit potential funding, opportunities, have local support from small island developing states (SIDS), and understand the context of these positions, while remaining conscious of the need for capacity building and true transformation possibilities.

Local Impact and Leadership

While the entire world is being affected by these global threats, these challenges also present more acute problems for coastal and island states. In response, several have offered vibrant leadership, especially in articulating how their exclusive economic zones (EEZs) can provide a platform for a more positive outcome in the years to come.

For instance, a group of Pacific island countries (PICs) joined together to raise $500 million to achieve carbon-free shipping in the Pacific Ocean by 2050.[19] The Pacific Blue Shipping Partnership, consisting of Fiji, the Marshall Islands, Samoa, Solomon Islands, Vanuatu, and Tuvalu, has established a 40% emission reduction goal for 2030 and a 100% decarbonization goal for 2050. These targets are expected to be achieved through various financial modalities such as grants from multilateral institutions, concessional loans, and direct private sector investment through regional "blue bonds." Alongside a $2 billion investment of sustainable and carbon-friendly technologies into the electricity and transport sectors under the Pacific Energy Summits since 2012, the Pacific Blue Shipping Partnership represents part of the enthusiasm of island states to invest in the future of their oceans.[20] In addition, The partnership also aims to acquire zero-emissions vessels and upgrade current passenger vessels with low-carbon technologies.[21]

Similar collaborative approaches were evident at the 2022 UN Ocean Conference held in Lisbon. As the conference materials highlighted, the Portuguese EEZ is 18 times the size of its land area. This is always one of the most useful early calculations when exploring the marine and maritime potential of any nation. The results are often most striking when considering SIDS, as the land-to-sea ratios of many of them are actually 99% sea within their EEZ. Hence, many commentators now much prefer to call them large ocean states. Indeed, that was the banner under which a number of PICs gathered at the Lisbon conference to present a united voice on ocean matters, pushing for greater international commitments and coordinated action.[22]

[19]   B. Doherty. 2019. Pacific Islands seek $500m to make ocean's shipping zero carbon. News release. *The Guardian*. 24 September.
[20]   Micronesian Center for Sustainable Transport. Pacific Blue Shipping Partnership.
[21]   Fiji's Ministry of Commerce, Trade, Tourism and Transport. 2020. *Decarbonising Domestic Shipping Industry: Pacific Blue Shipping Partnership*.
[22]   Secretariat of the Pacific Regional Environment Programme. 2022. *Pacific Large Ocean Island States Strategise for Amplified Pacific Voice at Second UN Ocean Conference*. Lisbon. 26 June.

That word "coordinated" is key. For too long, ocean management has suffered from a lack of coordinated planning and management, with competing or nonaligned interests or sectors suffering from a lack of comprehensive thinking and planning. Therefore, it was encouraging that a number of nations came together in a side event to call for integrated ocean management across the Pacific to "scale up innovative action for a healthy ocean, people and islands informed by science and traditional knowledge."[23] More of this type of leadership is needed.

Conclusion

The overexploitation of fisheries aside, our oceans are a largely untapped resource that contain enough energy to power and fuel our civilization, fight climate change, and continue to house and protect marine ecosystems.

While the facts related to issues such as coastal erosion, sea level rise, and marine ecosystem decline are incontrovertible, if we only view the oceans as a problem mainly associated with climate change, we overlook the fact that they can provide solutions to many of the world's problems. New opportunities can be realized if we break out of the problem narrative and develop a comprehensive ocean vision to truly understand the oceans and their strategic potential. The next chapter will explain and explore this new approach.

---

23   Pacific Community. 2022. *Sustainable Blue Pacific Continent: Scaling Up Action through Ocean Science and Traditional Knowledge for Informed Governance.* Side event of the UN Ocean Conference. Lisbon. 1 July.

CHAPTER 2

A New Blue Lens

In this chapter, we identify how governments and the private sector are using new approaches to promote a more synergistic and holistic management of ocean industries, marine conservation, and related issues. We will show how a positive outlook, along with sensitive and holistic stewardship of ocean resources, has promoted the continued emergence of the blue economy and sustainable long-term benefits.

What is the Blue Economy?

Box 1: Blue Economy Definition

The original term "Blue Economy" was proposed by the economist Gunter Pauli in 1994 in preparation for the 1997 Conference of the Parties (COP) 3 in Japan. It is a competitive business model that responds to the basic needs of all with what is locally available and that allows producers to offer the best at the lowest prices by introducing innovations that generate multiple benefits, not just increased profits. The concept and examples were given in Pauli's book of the same name.

Source: G. Pauli. 2010. *The Blue Economy: 10 Years, 100 Innovations, 100 Million Jobs.* Berlin: Konvergenta Publishing UG.

The blue economy concept has since evolved—from the term originally proposed by G. Pauli (Box 1)—and has been used by the World Bank to refer only to ocean issues, and that definition is used for this handbook. The concept recognizes that economic activities in the ocean, e.g., fishing, can affect the health of its ecological systems. Most of these activities rely on healthy marine ecosystems and all of them have the capacity to degrade ecosystem health.[24] The blue economy concept has led policy makers to develop policies that enhance and/or regenerate ocean health and to foster economic growth in a manner that is consistent with social equity and inclusion (footnote 24). The blue economy concept is also consistent with that of a circular economy, where material resources are designed to be utilized, reused, or recycled for as long as possible.[25]

Thus, the current concept of the blue economy acknowledges and respects the ecological boundaries that establish the safe operating space for economic activities as well as the minimum social standards that a fair economic model should address. It aims to allow socioeconomic development that does not cause environmental and ecosystem degradation.[26] Its economic framework is aligned with the UN SDGs, which

---

[24] P. G. Patil et al. 2016. *Toward a Blue Economy: A Promise for Sustainable Growth in the Caribbean.* Washington, DC: World Bank.

[25] Footnote 24; K. Raworth. 2017. *Doughnut Economics: Seven Ways to Think Like a 21st Century Economist.* London:. Penguin Random House; Global Factor. 2018. *4th Interim Delivery Assessment: Assessing the Existence of Key Elements of Blue and Circular Economy in the Caribbean.* Bilbao, Spain: Factor. 347 pp; and B. Bramley et al. 2021. *The Blue Economy in Practice – Raising Lives and Livelihoods.* Massachusetts: NLA International Ltd.

[26] S. Smith-Godfrey. 2016. *Defining the Blue Economy. Maritime Affairs: Journal of the National Maritime Foundation of India.* 12 (1). pp. 58–64; and World Bank. 2017. *The Potential of the Blue Economy: Increasing Long-Term Benefits of the Sustainable Use of Marine Resources for Small Island Developing States and Coastal Least Developed Countries.* Washington, DC.

enables people and nature to thrive in a balanced way, transparently acknowledging and addressing trade-offs based on the best available scientific evidence.

The ocean is being recognized as a significant potential source of economic recovery in the wake of the COVID-19 pandemic. Drivers of a growing ocean economy vary, which may include increasing familiarity with the ocean; new technologies to make it economically feasible to access ocean resources; and long-term growth and demographic trends, like increasing population, rising affluence, and demand for goods and services that increase the need for food security, alternative energy and mineral resources, seaborne trade, and coastal development.[27]

The blue economy approach does not resemble the former "growth at all costs" development model that has led to severe depletion of renewable natural resources, or a model where the ocean commons are effectively privatized to benefit a few.[28] Rather than just focusing on growth, under a blue economy approach, policy makers also focus on environmental and economic sustainability and benefits to broader society.[29] This approach has been the basis for the rapidly developing impact investment sector that now tops $1 trillion.[30]

Blue economic policy recognizes ocean-climate interactions and includes plans for the effects of climate change on coastal and marine resources (footnote 29). The aim is to increase the resilience of social-ecological systems to the impacts of climate change both now and in the future (footnote 29). Given the vulnerability to natural hazards of many developing member countries (DMCs) of the Asian Development Bank (ADB) and the increasing prevalence of some of these with climate disruption (e.g., more intense storms and rising sea levels), this is particularly pertinent to ADB's remit.[31]

The blue economy is thus a set of human activities that organize, in an integrated, equitable, and circular way, the production, distribution, exchange, and consumption of goods and services resulting from the exploitation of aquatic resources, or the use of the support provided by aquatic environments. These activities also contribute to improving the health of aquatic ecosystems by introducing protection and restoration measures. The blue economy is therefore most regularly articulated around the valorization of economic sectors and ecological components in a holistic manner. It suggests a new way of looking at the economic development of aquatic and marine ecosystems and the creation of jobs by referring to the principles of the circular economy. In this regard, the blue economy is in line with the logic of the UN 2030 Agenda and embraces all the SDGs, given its inclusive nature.

This alignment is echoed in ADB's interpretation of the blue economy under its *Action Plan for Healthy Oceans and Sustainable Blue Economies*. Along with a new ADB Oceans Financing initiative and Southeast Asia Financing Hub, it includes a call to transition away from an extractive model for the ocean economy and an evolution toward a longer-term ecologically, economically, and socially sustainable model. This prioritizes the protection and restoration of marine ecosystems and promotes inclusive livelihood opportunities, improving ocean health, and achieving SDG 14: Life Below Water. It ensures the health, resiliency, and food security of billions of people in the region. Its four pillars include the blue economy through sustainable ecotourism, fisheries, and aquaculture; ecosystem management; pollution control; and sustainable tourism.

---

[27]   The Economist Intelligence Unit. 2015. *The Blue Economy: Growth, Opportunity and a Sustainable Ocean Economy.* Briefing Paper for the World Ocean Summit 2015. Cascais, Portugal. 3–5 June.

[28]   B. Bramley et al. 2021. *The Blue Economy in Practice – Raising Lives and Livelihoods.* Massachusetts: Blue Economy pulse. NLA International; and G. Standing. 2022. *The Blue Commons: Rescuing the Economy of the Sea.* London. Pelican Books.

[29]   B. Bramley et al. 2021. *The Blue Economy in Practice – Raising Lives and Livelihoods.* Massachusetts: Blue Economy pulse. NLA International.

[30]   Y. S. Lamy, C. Leijonhufvud, and N. O'Donohoe. 2021. The Next 10 Years of Impact Investment. *Stanford Social Innovation Review.* 16 March.

[31]   Global Factor. 2018. 4th *Interim Delivery Assessment: Assessing the Existence of Key Elements of Blue and Circular Economy in the Caribbean.* Bilbao, Spain: Factor. 347 pp; and M. Witter. 2021. COVID-19: Intensifying the Existential Threat to the Caribbean. *Agrarian South: Journal of Political Economy.* 10 (1).

National Approaches to the Blue Economy

Creating a sustainable business model that utilizes the natural capital of the oceans to support people, the planet, and prosperity requires cooperation with global stakeholders including international institutions, governments (national to local), civil society, and corporations. It is a genuinely achievable approach and many countries around the world are championing the model.

At the national level, the potential complexity of blue governance should not be underestimated. It involves overseeing several types of marine activities that may be professionally managed but have not been fully integrated under a unifying strategy or governance framework.

For example, integrated blue governance will potentially involve closer coordination across a range of agencies; more integrated planning activities, including maritime spatial planning; new forms of monitoring, including potentially novel forms of blue accounting; and integrating, where appropriate, relevant blue international standards such as Blue Flag for tourism. A coherent, fit-for-purpose blue governance framework is thus required to build a multiscale organizational scheme and foster harmonious long-term collaboration among local, national, and regional institutions.

The European Union (EU) has also taken many steps to formalize its recognition of and desire to improve the blue economies of Europe and EU member states. Designed to achieve the objectives of the European Green Deal, the European Commission is proposing a new approach for a sustainable blue economy. It focuses on carbon neutrality, switching to a circular economy, preserving biodiversity, supporting climate and coastal resilience, ensuring sustainable food production, and strengthening coastal and marine management. The European Investment Bank and European Commission will increase their cooperation with member states to achieve a sustainable blue economy by 2027.[32]

India, an economic powerhouse and one of the fastest developing countries in the world, has a coastline of 7,500 kilometers (km), 12 major and 200 minor ports, and recognizes that 95% of the country's business is supported by the blue economy and maritime transportation.[33] India has been developing a Blue Economy Policy Framework since 2020 through its Maritime and Earth Sciences Ministries.[34] This has involved further integration in the form of India's nascent Blue Economy Coordination Committee.[35] Such an approach is also being pursued in other South Asian nations such as Pakistan, which is developing an integrated Blue Economy Roadmap, and Bangladesh, which has initiated a Blue Economy Cell to coordinate thinking and implementation.[36] Each relevant institution holds a piece of the blue economy jigsaw. The challenge for governments is to provide a unified vision.

This coordinated approach is not just at the national level. Already, we are seeing how integrated blue economy thinking is trickling down to state and, in some cases, city level. In March 2022, the city of Umm Al Quwain, the capital and largest city of the Emirate of Umm Al Quwain in the United Arab Emirates, launched its own Sustainable Blue Economy Strategy 2031.[37] The strategy balances economic growth with environmental stability, aiming to double gross domestic product (GDP) by 2031 and for the blue economy to contribute 40% of that total. A net-zero emissions target has also been established for 2031, by which time a total of 20% of

[32]　European Commission. 2021. European Green Deal: Developing a Sustainable Blue Economy in the European Union. Press release. 17 May.

[33]　M. Juneja. 2021. Blue Economy: An Ocean of Livelihood Opportunities in India. *The Energy and Resources Institute.* 12 March.

[34]　Economic Advisory Council to the Prime Minister Government of India. 2020. *India's Blue Economy: A Draft Policy Framework*. New Delhi.

[35]　A. Rahman Rasel. 2016. Govt Forms Blue Economy Cell. *Dhaka Tribune.* 3 December.

[36]　Netherlands for the World Bank. 2020. eC2: Analysis of Blue Economy Development in Pakistan and Developing Blue Economy Development Roadmap.

[37]　*MENAFN-Khaleej Times.* 2022. Umm Al Quwain launches its Sustainable Blue Economy Strategy. 29 March.

Umm Al Quwain is set to be dedicated to nature reserves. All these actions combine under the banner of transforming Umm Al Quwain into the "capital of the blue economy."

Crucially, the blue economy concept is moving beyond the theoretical stage, as governments are identifying ways in which they need to modify existing practice and governance arrangements to establish a clearer picture of blue value. This is perhaps best exemplified by the US government, which in June 2022 revealed that its blue economy sectors contributed $361 billion to GDP in 2020. Such analysis was made possible because of the government's adoption of satellite accounting methods for blue economy activities. On release of the figures, a government spokeswoman said: "These numbers are stark proof of just how tied the Blue Economy is to the prosperity of the rest of our nation. What happens in this sector has ripple effects through the entire nation, which relies on [the] Blue Economy as a driver of jobs, innovation, and economic growth."[38]

Around the world, examples of blue governance can be seen to reveal two fundamental lessons learned. First, that economic development must go hand in hand with carefully planned and managed regenerative activities. Second, that new blue governance structures must be introduced that enable enhanced coordination and integration across all relevant ministries, sectors, and broader stakeholders.

These two "lessons learned" form the basis of the Marine Aquaculture, Reefs, Renewable Energy, and Ecotourism for Ecosystem Services (MARES) approach. Development projects can be seen as a logical extension of this new approach to the blue economy at an operational level. And where possible, MRE and sourced non-fossil fuel will play an important role.

Further Enabling Blue Development

Harnessing the blue economy encapsulates a holistic ecosystem of investment into the business of the oceans. This is done through viewing the oceans as an opportunity, rather than just a crisis to be mitigated. In addition to new strategies and governance arrangements, attention must be paid to the investment framework.

As well as planning to invest $5 billion through its Healthy Oceans Action Plan, ADB is developing thought leadership and knowledge products to provide an accessible framework to guide thinking in this respect.[39] In a series of articles in 2021,[40] Ingrid van Wees, ADB's then vice-president for Finance and Risk Management, put forward specific elements that are necessary to have in place to develop the blue economy, globally, regionally, and in individual nations. These included

- **Adopting holistic strategies.** Approaches must be broad and systematic as the health of the world's oceans is a crosscutting issue that does not just affect a single industry, sector, or country—it is a threat to the entire planet.
- **Natural resource accounting.** The goods and services the ocean provides are often undervalued. The inclusion of natural assets in public sector balance sheets will increase transparency, improve the value of blue assets, and reinforce the need to protect these vital resources for the next generations.

38   The WorkBoat. 2022. 'Blue economy' contributed $361 billion to GDP in 2020, Department of Commerce says. 9 June.
39   ADB. 2019. ADB Launches $5 Billion Healthy Oceans Action Plan. News release. 2 May.
40   ADB. 2021. Toward a Blue Deal to Restore the World's Oceans - Ingrid van Wees. Op-ed/Opinion. 15 June.

- **Sound investment ecosystems.** Stronger enabling environments are required to encourage blue economy investments. This could include more focused governance arrangements, incentivizing ocean-positive businesses through taxes and subsidies, and discouraging or disallowing ocean-impacting businesses through fiscal and regulatory reforms.
- **Channeling private sector finance.** While sovereign investments remain essential, governments alone cannot be expected to have access to the scale of capital required to lead transformational change in the seas and oceans. The private sector can fill the gap, though this may involve developing new business models and careful consideration of what may involve the monetization of what is currently a shared common resource.

To achieve a thriving blue economy, it is useful to align strategic thinking and suggestions to areas of policy and development already well understood by key stakeholders. One recent study in Indonesia showed policy makers how, for example, they can support maritime businesses in coastal states to implement blue economy principles and achieve all relevant SDGs at the same time.[41] As the SDG agenda has already been integrated into planning and activities in many government agencies, it can act as a useful bridge to explain and embed blue economy thinking.

Such developments described in this chapter—when supported by an enabling environment and aligned with the kind of innovative new business models such as the potential export of hydrogen within the renewables market—could encourage a new wave of private investment to maximize government interventions. Less immediately "bankable" projects (i.e., less well-designed for financing modalities) will need specially targeted assistance through blended finance that can catalyze opportunities by lowering risk and enhancing profitability to attract private sector participation and capital.

Finding the synergies and funding modalities suited to each subproject requires project teams to

(i)   assess the conditions of the market and quantify potential natural capital and socioeconomic gains, while promoting long-term ecological sustainability, ocean health, and financial prosperity;

(ii)   develop selection criteria and advise on helpful policy framework developments to create financial instruments and products;

(iii)   mobilize concessional finance, where necessary and where clear returns are apparent; and

(iv)   prepare well-designed project pipelines that will realize opportunities by considering available financing modalities.

Embedding Key Principles

Despite its promise, a word of warning must be sounded about the blue economy. There are a number of definitions of the blue economy in the literature and various states and other stakeholders that have very different operational interpretations. Many nations continue to place high priority on the economic gains in the ocean economy with little or no consideration for social inclusion or environmental sustainability.[42] For example, in Africa, there are large-scale ocean economy projects that have marginalized local communities and severely damaged the environment (see examples in footnote 42). In these examples, top-down management of coastal

---

[41]   K. Akhir et al. 2021. Designing and Mapping the Blue Economy Company Index (BECdex) to the Sustainable Development Goals (SDGs) for Maritime Companies in the Coastal States. *International Journal of Innovative Science and Research Technology.* 6 (9).

[42]   I. Okafor-Yarwood et al. 2020. The Blue Economy–Cultural Livelihood–Ecosystem Conservation Triangle: The African Experience. *Frontiers in Marine Science.* 7.

infrastructure projects has disregarded the needs of local populations and led to the destruction of the very coastal ecosystems on which those communities rely for their subsistence (footnote 42).

New economic models are now required to encourage policy making that directs investment to initiatives that regenerate marine and coastal ecosystems under a blue economy umbrella and increase support to communities whose basic needs for nutritious food, energy, and good working conditions have not been met.[43]

In NLA International's recent report, The Blue Economy in Practice (footnote 29), five key principles were put forward that if embraced would enable people and the planet to thrive under a blue economy approach. The report suggests that blue economies should be regenerative, adaptive, inclusive, sustainable, and evidence-led:

Regenerative. To support a path toward greater resilience to changing climate, ambitious conservation and restoration measures, including protecting at least 30% of the ocean via networks of effective marine protected areas, are necessary. The "core" planetary boundaries such as climate change and biosphere integrity are fundamental to a stable earth system. We have exceeded both boundaries,[44] along with unsustainable land use and biochemical flows (e.g., overuse of nitrogen and phosphorous).

The goals to halt and reverse the trends in climate heating and biodiversity loss are urgent and these can be met by protecting more of the ocean. The protection and regeneration of the ocean should be taken holistically by combining fully protected areas with sustainable fisheries management and undertaking measures to safeguard migratory species, which are essential to the function and health of the ecosystem.

Adaptive. Climate change is causing increases in ocean temperatures that are killing coral reefs, ocean acidification, sea level rise, and increased wind speeds, wave heights, and storm surges during hurricanes or typhoons. The ocean's influence on climate is significant and vice versa. Consequently, these caused changes in ecosystem function as well as the distribution and extent of marine species. More and larger marine protected areas, restoration of coastal blue carbon ecosystems, and seaweed farming are some of the ocean-climate solutions that should be prioritized in national "net-zero" mitigation and adaptation agenda and blue economy strategies. These solutions provide triple benefits for people, planet, and prosperity. Plans relating to coastal and marine management should be regularly updated based on the best available scientific knowledge of the ocean-climate nexus and be adapted to the latest conditions and predictions.

Inclusive. To secure holistic and balanced decision-making processes, new approaches to ocean governance, policy, and management are necessary. These approaches should be inclusive, gender-balanced, and participatory rather than top-down. Biodiversity loss and climate change are systemic problems that require systemic solutions. To generate fresh insights and bottom-up innovation for policy initiatives, it is critical to empower local communities and invite wider sectors of society such as entrepreneurs, commerce, academia, nongovernment organizations (NGOs), and impact investors to participate in solutions. These should seek to address current inequities and conflicts in accessing and gaining value from ocean resources and encourage joint management of coastal ecosystems, for instance.[45]

---

43 U. R. Sumaila et al. 2021. Financing a Sustainable Ocean Economy. *Nature Communications*. 12 (1). 3259.
44 W. Steffen et al. 2015. Planetary Boundaries: Guiding Human Development on a Changing Planet. *Science*. 347 (6223).
45 N. Bax et al. 2021. Ocean Resource Use: Building the Coastal Blue Economy. *Reviews in Fish Biology and Fisheries*. 32 (1). pp. 189–207.

Sustainable. It is necessary to put in place policy measures to stop the overexploitation of ocean resources, particularly marine fisheries. More than a third of assessed fish stocks are at biologically unsustainable levels (footnote 2) and overfishing is the primary cause of a 71% reduction in shark population since 1970, such that 75% of shark species are threatened with extinction.[46] Generally, the main factors contributing to biodiversity loss in the ocean are overfishing and the destructive effects of fishing.[47] Based on estimates, the capture of large fish from 1950 to 2014 has prevented the sequestration of about 22 million tonnes of carbon from the sinking fish carcasses that would have otherwise fallen to the seabed. This is in addition to direct emissions from fishing over the period of 730 million tonnes of carbon dioxide.[48]

To lessen the environmental harm caused by activities like fishing, e.g., by catch-reduction measures, technical measures can be introduced (footnote 47). Support should be provided to wasting less fish and seafood as more than a third is currently lost (footnote 2). The circular economy principles should be used to stop the flow of waste into the ocean and recover new economic value from waste materials. A precautionary approach to new ocean-based activities should be adopted with full consideration on the effects on climate and biodiversity as well as circular economy alternatives in decision-making.

Evidence-led. Decisions should be undertaken considering the user needs and a foundation of the best available natural and social science, data, and existing knowledge, including indigenous knowledge. To measure the success of a blue economy approach, new holistic measures of progress are necessary aside from GDP, which reflect ocean health and wealth, as well as associated human well-being and prosperity in ocean and national accounts. Policy makers and their implementing organizations should monitor these values over time and adjust policies as necessary to ensure that nature and society are thriving in a more balanced way and that the needs of future generations can be met.

These principles are in various ways reflected in the ethos of the MARES project.

Conclusion

The blue economy's new strategic framework promotes a more synergistic, holistic approach to the governance, protection, and management of ocean industries and marine conservation. This allows for the proactive planning of sustainable long-term impacts and, where evident, should also provide confidence in investor networks that the right enabling environment is in place to encourage innovation. In the next chapter, we look at some blue sectors of particular interest to the MARES project and highlight some interesting innovations in each.

[46]   N. Pacoureau et al. 2021. Half a Century of Global Decline in Oceanic Sharks and Rays. *Nature*. 589 (7843). pp. 567–571.
[47]   A. Rogers et al. 2020. *Critical Habitats and Biodiversity: Inventory, Thresholds and Governance*. Washington, DC: World Resources Institute.
[48]   G. Mariani et al. 2020. Let More Big Fish Sink: Fisheries Prevent Blue Carbon Sequestration—Half in Unprofitable Areas. *Science Advances*. 6 (44).

CHAPTER 3

Investment in Blue Economy Innovation

This chapter highlights how increased confidence in the potential of the blue economy is driving new levels of investment, exemplified by innovations in three specific sector interests of the Marine Aquaculture, Reefs, Renewable Energy, and Ecotourism for Ecosystem Services (MARES) project.

Introduction

While the blue economy concept has evolved over 30 years, it is getting a lot of positive attention now from governments and the private sector as a sustainable pathway. This is reflected in increased and increasingly confident investment in specific marine sectors and in technological innovation.

Some examples are given that show how this investment is supporting innovative approaches in the three key sectors initially identified for support in the MARES project to align with marine renewable energy (MRE) options: marine aquaculture, cultivated reefs and other nature-based defenses, and marine ecotourism.

Each of these developing sectors offers new opportunities for island and coastal states. The value and the potential for investing in aligned multifunction MARES projects will be examined in later chapters.

Marine Aquaculture

Introduction to the Sector

The "M" of MARES stands for marine aquaculture or mariculture and this term includes seaweed, shellfish such as clams, and fish. Wild-caught fisheries have been overfished and and, despite improvements in technology and thousands more boats, the global catch has been flat since about 1985. In contrast, global aquaculture production nearly tripled in volume between 1995 and 2007, and has continued to grow rapidly ever since, propelled by the public's demand for seafood and the omega-3 fatty acids found in oily fish, believed to be good for the heart.[49] The mariculture sector will continue to grow in importance in the coming decades. The People's Republic of China (PRC) is the largest producer of seafood in the world and aquaculture production (freshwater and marine) in the PRC is more than half of global aquaculture production. The PRC's aquaculture production overtook that of its wild-caught fisheries in about 2009.

By 2016, global production from aquaculture had reached 80 million tonnes, providing 53% of all fish consumed by humans as food.[50] The World Bank estimates that the size and scale of the aquaculture market will continue to

---

49   R. L. Naylor et al. 2009. Feeding Aquaculture in an Era of Finite Resources. *Proceedings of the National Academy of Sciences.* 106 (36).
50   FAO. 2018. Is the planet approaching "peak fish"? Not so fast, study says. 9 July.

grow until 2030, by when it will account for 62% of all the seafood we consume.[51] This is not only important news for the blue economy—FAO has also pointed out that aquaculture production continues to outpace every other food sector, growing at an annual compound rate of nearly 6% since 2010.[52] In 2016 alone, the aquaculture sector increased production by some 4 million tonnes year on year. Additionally, in its comprehensive 2016 Ocean Economy report (footnote 14), the OECD highlighted the industrial marine aquaculture market as one of the key prospects for high long-term growth, with employment within the sector predicted to increase by over 150% by 2030.

Many countries are already heavily reliant on aquaculture. Like the PRC, the industry accounts for over 50% of Philippine fisheries output (in contrast to EU markets, where aquaculture accounts for approximately 25% of overall output) and is regularly the one market sector whose steady growth allows the country overall fisheries output statistics to retain some buoyancy, as standard commercial and municipal capture fisheries can fluctuate significantly.[53] Fish makes up more than 50% of the animal protein consumed in the Philippines; hence, aquaculture plays an important role in supplying the most basic nutrition a population that nevertheless experiences substantial involuntary hunger from time to time[54]—and this before the shock of COVID-19 and the recent global cost-of-living crisis.

With such compelling statistics, it is no surprise that countries and businesses around the world are taking great interest in how they can support and benefit from an industry that is universally accepted to offer high growth in the coming decades and help to counter the potential for hunger in a wide range of diverse communities.

In Southeast Asia, Viet Nam's authorities are embracing the opportunity as they recently revealed the country's ambitious intentions to become the world's top aquaculture producer,[55] with a headline aim of achieving aquaculture export value of $8.86 billion by 2050. Achieving top spot in the global aquaculture rankings will mean overtaking the current global top three, which are the PRC, Indonesia, and India. Such ambition cannot be achieved without significant investment, and fisheries leaders are certainly doing so. Taking one example, the Vietnam Association of Seafood Exporter and Producers announced in 2018 that the central coastal province of Phu Yen will invest nearly D2.12 trillion ($85.32 million) to help develop its aquaculture industry up to 2025.

Like many southeast Asian countries, the majority of its recorded aquaculture output actually comes from small-scale independent fish farmers—in Viet Nam's case, involving some 50,000 households. While these producers tend to use traditional methods and basic equipment, the industrial aquaculture race is also driving demand for new and innovative technology-led practices.

Aquaculture Innovations

None of these eye-grabbing targets will be achieved with traditional methods alone, and innovative applications across all elements of aquaculture development and delivery are beginning to attract investment.

Achieving ambitious growth targets within the aquaculture industry will require more accurate monitoring to maintain stock health, and therefore value, and a better understanding of the environmental impacts to ensure the industry can grow sustainably.

51   World Bank. 2014. *FISH TO 2030: Prospects for Fisheries and Aquaculture*. Washington, DC.
52   *US Soybean Export Council*. 2018. Aquaculture is Fastest Growing Food Production Sector, According to FAO Report. 16 July.
53   Philippine Statistics Authority. 2019. Fisheries Situation Report for Major Species: January to December 2021.
54   G. K. Cabico. 2018. SWS: 3.6M Filipino Families Experienced Hunger in Q4 2017. *Philstar Global*. 22 January.
55   T. Dao. 2018. Vietnam Poised to Become Top Player in Ocean Aquaculture. *Seafood Source*.

New technologies are helping aquaculture development, for example, by determining the best locations for aquaculture facilities. As with many blue economy sectors, satellite-enabled systems are beginning to offer value here. As an example, the United Kingdom (UK) company, TCarta, delivered a ground-breaking project in the Arabian Gulf in 2018, as they put their satellite-derived bathymetry data to use. BMT, an innovative UK engineering and scientific consultancy firm, as part of the service they are providing to the Abu Dhabi Environment Agency, used specialized TCarta data sets to help in the selection of new fish farming sites in the area.[56] Companies that collect bathymetry from satellites have been expanding their databases in recent years, both to address the significant lack of accurate seafloor depth measurements globally, and to take advantage of the new commercial and environmental opportunities provided by such high-resolution satellite imagery. In this instance, hydrographic modeling software helped to identify ideal fish farming sites based on water depth thresholds (to accommodate fish cages) and alignment with natural subsurface channels whose water currents can wash away waste.

Environmental impacts are a major challenge for aquaculture development. There are many issues—having the right environmental conditions for fish, shellfish, or algae to thrive; being able to monitor those conditions to ensure they remain stable and do not deteriorate and threaten the health of the cultivated organisms; and finally, making sure that the aquaculture site does not harm the surrounding environment.

Taking the middle of those issues, aquaculture managers must guard against outbreaks of disease that can threaten their stocks. Healthy fish require monitoring. Responsible food producers must adhere to strict environmental and health regulations. Aquaculture requires careful consideration of several important factors, including animal health, hygiene, and particularly water use. Disease and sickness can spread easily among the animals if not discovered swiftly; and, because of the design of these aquatic farms, outbreaks can also spread to wild populations, putting sites at danger of immediate closure.

Fish farmers in Lake Toba, the biggest lake in Indonesia, have suffered more than most in recent years in relation to these issues, with two separate incidents (one in 2016 and one in 2018), devastating their operations and leaving millions of fish dead on each occasion.[57] Perhaps the most worrying aspect here is that both incidents came with almost no warning. While some fish farmers in one of the incidents reported that they had noticed some unusual activity within their floating nets, for many, it was completely unanticipated, giving them no time at all to respond with mitigating measures.

These examples were attributed to a sudden depletion of oxygen in the water. Such conditions can come about due to a toxic mix of pollutants building up in the lake, unfavorable weather conditions, and poor practice from local fish farmers whose operations reduce the water quality. Not being able to sustain such environmental and economic shocks on a regular basis, the Indonesian authorities are taking measures to help counter the possibility of such incidents happening again. Embracing this, in October 2018, fisheries officials launched a predictive calendar to alert fish farmers in Lake Toba to the emergence of water conditions that may prove threatening to their fish stocks.[58] The alerts are intended to ensure that fish farmers react positively to worrying signs of water health by, for example, reducing feed amounts that can in turn exacerbate emerging problems.

However, such planning tools (sensible as they are and even if available online) are just one of the more basic options available. Others are utilizing the power of machine learning to make a difference. For instance, the aquaculture health response system being established by analytics firm, Manolin, gathers large amounts of data

56   *DM Insights on Location.* 2018. Satellite Derived Bathymetry from TCarta Plays Key Role in Aquaculture Siting Project. 19 February.

57   A. Karokaro. 2018. Another Mass Fish Kill Hits Indonesia's Largest Lake. *Mongabay Environmental News.* 24 August.

58   B. Gokkon. 2018. Indonesian Fish Farmers Get Early-Warning System for Lake Pollution. *Mongabay Environmental News.* 25 September.

and notifies fish farmers at once when a parasitic sea lice outbreak appears to be spreading in their region.[59] As with any Big Data project, there are only a number of issues that a unified system can address. There is too much data to make proper sense of; it can be difficult to combine data from multiple sources on multiple systems; much of the data is actually on paper; etc.

The Manolin team, based in Bergen, Norway, is developing an automated system to get key information to interested operators much more efficiently. They may have a willing market when it has been estimated that sea lice now impact 17% of all aquaculture production costs, an impact that has grown 233% since 2010. In Norway, the annual loss of stock caused by sea lice is estimated to be $575 million (£500 million). It is worth considering the current manner of operations to better understand the potentially transformative effect of services, such as those offered by Manolin. Salmon farmers are obliged to undertake weekly surveys of their cages to measure sea lice levels and report findings to a central authority who collects the data. In many cases, the current method is to manually check 10 fish in a cage of circa 200,000 specimens. This rate of sampling does not represent a statistically significant sample, so serious vulnerabilities remain. New automated options provide much more effective and comprehensive methods of monitoring.

Many cultured species of fish are carnivorous and historically have been grown on manufactured feed made partly from ground up by-catch or "trash fish." A major cost of fish culture is the cost of imported feed. A major advance for fish aquaculture has occurred in the past 2 years, with the push to develop fish food that is made from algae that includes oils similar to omega-3. The technology of cultivating various strains of algae has also advanced to the stage where it is possible to purchase a mobile laboratory designed specifically to grow algae.

More innovations will also be required to help marine aquaculture to adopt safeguards that ensure that the industry minimizes environmental harm—including advanced environmental monitoring options. This also relates to decision-making to ensure that, for instance, aquaculture does not lead to mangrove conversion, or lead to environmental degradation issues from abandoned aquaculture farms and platforms.

Cultivated Reefs

Introduction to the Sector

The promise of cultivated reefs is another area of early interest for the MARES project, as well as other approaches to restoring and fortifying nature-based defenses.

The definition of ecosystem restoration is "to return an ecosystem to a close approximation of its condition prior to disturbance."[60] Such restoration must restore the structure and function of the affected ecosystem and integrate it into the wider seascape.[61] The first thing to point out with respect to coral reef restoration is that it can only be effective if the causes of reef degradation are known and have been reduced or removed (footnote 61). For example, if water quality is very poor and is leading to loss of reef-forming corals then the causes of poor water quality must be addressed before restoration can take place (footnote 61). A second point is that coral reefs, particularly in the Indo-Pacific, may include hundreds of species of corals, thousands of species of fish, shellfish, other invertebrates, and algae. Therefore, the ideal goal of "restoring" a reef following serious damage is not realistic. However, given a few years, rehabilitation to a functioning coral reef is possible.

---

59   R. Fletcher. 2018. Fish and Chips: How Computer Analytics Can Transform Aquaculture. *The Fish Site. 17 September.*

60   W. Precht and M. Robbart. 2006. Coral Reef Restoration. In W. F. Precht, ed. *Coral Reef Restoration Handbook.* CRC Press.

61   Edwards, A. J., ed. 2010. *Reef Rehabilitation Manual.* St. Lucia, Australia: Coral Reef Targeted Research & Capacity Building for Management Program.

Therefore, the first step in any reef rehabilitation project is to establish the local, regional, or even global processes that have impacted a reef and caused its degradation and to assess whether they can be mitigated.

Outside of specific incidents that damage coral reef ecosystems such as ship groundings or impacts from specific coastal development, reefs are often degraded by a combination of local, regional, and global stressors. In this case, it is often the aim to restore reefs to as close as possible to the natural state usually identified either from nearby reef systems in good health status or to examine quaternary (subfossil) reefs in the same locality to assess what was present previously. The overall aim of such rehabilitation is to restore ecosystem functions, aesthetic values, and services so that the local people can continue to operate (e.g., fishing or carrying out tourist activities) and that local cultural requirements are met.

Approaches to reef restoration can be summarized as follows: indirect action (no active restoration), reef repair, artificial reefs, and cultivated reef innovations.[62] These approaches vary in cost and technological approach and have to be specifically tailored to restoration objectives and the locality.

Indirect Action

The simplest approach to restoring a coral reef is to remove the stressors that are degrading it (footnote 62). Preventing overfishing of grazing fish that consume algae, improving water quality by treating sewage, preventing sedimentation from coastal development, or restoring associated ecosystems (e.g., mangrove forests or seagrass beds) can all improve the natural settlement of young corals and growth of coral reefs.

Reef Repair

Reef repair can involve emergency repair of a reef following an acute disturbance (e.g., ship grounding or hurricane) or specific interventions to restore the structure and topographic complexity of a reef (footnote 62). Emergency repair can involve finding dislodged corals or coral fragments and reattaching them to a reef using various methods before they die or placing them in safe environment (e.g., a coral nursery) until they can reattach to the reef (footnote 62).

Artificial Reefs

Artificial reefs aim to provide a stable fixed substratum to which corals may recruit and grow.[63] These structures may be formed in a way to create three-dimensional complexity, which can also attract fish and other reef-dwelling organisms (footnote 63). A variety of materials can be used to create artificial reefs including rocks, limestone rocks, cement, or combinations of these. Cement structures can be textured and formed into a variety of complex shapes, e.g., Reef Balls, EcoReefs, Grouper Ghettoes, A-Jacks, Warren Modules, and Department of Environmental Resources Management Modules (footnote 63).

Another approach is to form a complex shape from conductive metal bars or a frame and to pass a current through it to cause electrolytic deposition of limestone onto the metal. This is commonly known as mineral accretion. Corals can then be transplanted and attached to the limestone or larvae may naturally attach to the limestones-coated structures.[64]

---

[62]   M. Y. Hein et al. 2020. *Coral Reef Restoration as a Strategy to Improve Ecosystem Services – A Guide to Coral Restoration Methods.* Nairobi: UNEP. 60 pp.

[63]   L. Kaufman. 2006. If You Build It, Will They Come? Toward a Concrete Basis for Coral Reef Gardening. In W. F. Precht, ed. *Coral Reef Restoration Handbook.* CRC Press; and B. Zimmer. 2006. Coral Reef Restoration: An Overview. In W. F. Precht, ed. *Coral Reef Restoration Handbook.* CRC Press.

[64]   T. Takagi. 2019. Energy Production: Biomass – Marine. In M. Ueda, ed. *Yeast Cell Surface Engineering: Biological Mechanisms and Practical Applications.* Springer Singapore. pp. 29–41.

Cultivated Reef Innovations

Many interesting innovations to protect, preserve, or enhance coral reefs are being used to rehabilitate damaged coral reefs.

Coral gardening is a low-cost but effective approach being deployed in many parts of the world to preserve and regenerate coral. Various types of algae live on healthy coral reefs. However, some types can grow quite large (macro algae) and compete for space on the reef with corals. Normally, certain herbivorous fish and shellfish eat the algae allowing corals to thrive. In cases where there are lots of nutrients (fertilizer) in the water, this helps the algae to grow and out-compete corals. Therefore, weeding—removing large algae from reefs—can improve the health of the reef by allowing corals to grow. This process has been applied in many areas, including the Great Barrier Reef, where it led to noticeable improvement in coral regeneration.[65] With the basic process proven to be effective, it will be interesting to observe if any new technologies or innovative approaches can be developed to increase speed of deployment.

One of the most effective examples of coral restoration in Belize has managed to plant hundreds of thousands of coral fragments, to rebuild fragile ecosystems ravaged by hurricanes.[66] The key learning from this project points to the power of community involvement. Local guides, fisherfolk, divers, and snorkelers are all trained to plant coral and subsequently monitor its development. Over a decade of activity, this community-focused system has increased coral coverage in protected areas off the coast of Placencia from 6% to 60%, making it one of the most successful and enduring coral regeneration sites in the world.

The effects of this success will be broad, as it has been estimated that the Belize Barrier Reef accounts for some 15% of the country's GDP and provides work for approximately 200,000 people in the fishing and tourism industries.[67]

Taking a more technological approach, a collection of environmentalists in French Polynesia are using artificial intelligence (AI) to help repair stressed coral reefs. The team identifies groups of more resilient super corals in their target sites, cutting and moving a small portion of them to a nursery site where they mature for a year and then replanted onto degraded areas of the reef. AI comes in by monitoring the growing process and providing important information on the ecosystem, e.g., by using a 360-degree camera that identifies and labels fish species as they return to the reef. The technology integration also allows interested supporters to watch a livestream of reef activity.[68]

AI has also been trialed recently for coral conservation in Indonesia. Researchers were able to photograph a distance up to 1.5 miles (or 2 kilometers) in a single dive through the 360-degree cameras fitted on underwater scooters.[69] During a period of "supervised learning," the AI recognition software begins to identify corals and other formations of interest by using algorithms and its own judgment. The amount of collected images were subsequently analyzed more quickly than human scientists possibly could, reducing the individual image analysis time from up to 15 minutes to just a matter of seconds. This made it possible to quickly assess the effect of global-warming-induced coral bleaching to the study area between 2014 and 2017. As more coral bleaching occurs globally, such major cost- and time-saving AI-related approaches can only be welcomed.

[65] S. Gibbens. 2022. Why Scientists Are 'Weeding' Coral Reefs. *National Geographic.* 24 February.
[66] R. Collett. 2022. Why is the Belize Barrier Reef One of the Most Successful Coral Restoration Projects in the World? *Euronews.green.* 8 May.
[67] World Wildlife Fund for Nature. Saving Belize's Magnificent and Endangered Barrier Reef.
[68] Coral Gardeners.
[69] L. Rosen. 2018. 360-Degree Underwater Cameras and Artificial Intelligence May Save Coral Reefs. *21st Century Tech Blog.* 16 August.

On the autonomy front, the RangerBot becomes a new adversary of millions of coral-eating crown-of-thorns starfish (COTS) in the Great Barrier Reef. A roboticist from Queensland University of Technology has created this autonomous underwater vessel that not only can find COTS, but because of recent, separate academic progress in understanding lethal threats to the starfish, can actually kill any COTS it encounters with a single injection of a derivative of bile.[70] This powerful combination of autonomy, recognition, and robotics shows how such confluence of technologies bodes well for environmental conservation.

One new tool showing promise of the required scalability is related to the process where marine scientists halt reef degradation by "sowing" corals. Traditionally, this has been a time-consuming, expensive, and laborious process (capturing coral seed during yearly mass spawning events)—growing new corals in the lab and attaching them by hand to existing, denuded coral. Ohio-based nonprofit Secore International is one company leading the way in developing this approach. They cultivate the coral larvae in specially designed star-shaped concrete "seeding units" that allow them to simply be "wedged" into existing reefs, rather than having to spend much more time attaching them manually. The company estimates that this process, likened by some to farmers scattering seed on land, could reduce processing time by over 90%, thus making it a more viable option for mass-scale adoption.[71]

A wide range of artificial reef projects continue to flourish around the world. For example, reefs off the North Carolina coast are formed by donated concrete pipes damaged by hurricanes, steel hulls from old tugboats, and bridge demolition debris.[72]

It is encouraging to note that as they mature, artificial reef projects can also demonstrate the value they are adding. For example, the Malaysian state of Sarawak has been investing in what is dubbed the longest artificial reef in the world, with nearly 17,000 artificial "reef balls" being placed along 1,000 km of its coastline. The reef balls are made of concrete about 1 meter (m) high, with large holes that allow fish to pass through. In a short amount of time, it has been estimated that local fishermen had seen their income increase more than three-fold because, it was claimed, of the enhanced marine environment.[73] Unfortunately, the reef balls are so lightweight that they can be pushed by waves and end up on the beach during storms, and only a few coral species will settle on concrete surfaces, so corals usually need to be attached by hand.

Eco- and Nature-Based Tourism

Introduction to the Sector

Finally, the MARES project is looking to find ways to integrate nature-based and ecotourism initiatives into its mix of MRE and other blue economy projects.

Nature-based tourism is the umbrella product where the main motivation is the enjoyment and recreation within the "natural" space. Experts have defined it as combining all forms of tourism where relatively undisturbed natural environments form the primary attraction or setting.[74] It can include activities that are consumptive

70   J. Adams. 2021. Killer Robot Hunts Down Crown of Thorns Starfish. *Reefbuilders*. 13 May.
71   Secore International.
72   J. Jurney. 2022. NCDOT Providing Damaged Concrete Pipes to Help Artificial Reefs off the Coast. *WUNC North Carolina Public Radio*. 1 March.
73   G. Pei. 2020. Sarawak to Have Longest Artificial Reef in the World. *Free Malaysia Today*. 26 October.
74   R. C. Buckley. 2009. *Ecotourism: Principles and Practices*. Wallingford: CAB International; and D. Newsome, S. A. Moore, and R. K. Dowling. 2002. *Natural Area Tourism: Ecology, Impacts and Management*. Clevedon: Channel View Publications.

and adventurous as well as nonconsumptive and contemplative, which in turn can include ecotourism[75] and conservation tourism.[76] Under this premise, the nature tourism subproducts of ecotourism, adventure tourism, cultural tourism, and nautical tourism can be directly associated with the tourism offerings of island nations.

Ecotourism

Ecotourism can be defined as "responsible travel to natural areas that conserve the environment, sustains the well-being of the local people, and involves interpretation and education" according to The International Ecotourism Society. Additionally, the UN World Tourism Organization refers to ecotourism as having the following characteristics:

- All nature-based forms of tourism in which the tourists are primarily drawn there to observe and appreciate nature as well as the local cultures prevailing in natural areas.
- It has interpretation and educational elements.
- It is typically, but not exclusively, organized by specialized tour operators catering to small groups. The service provider partners at the locations tend to be small, locally owned businesses.
- It reduces adverse effects on the natural and sociocultural environment.
- It encourages the preservation of natural areas that are used as ecotourism destinations by
    - generating economic gains for host communities, organizations, and authorities in charge of conserving natural areas;
    - providing alternative work opportunities for local communities; and
    - raising awareness on the need to preserve natural and cultural resources, both among residents and tourists.

The size of the global ecotourism market was valued at \$181.1 billion in 2019, and it is expected to grow at a 14.3% compound annual growth rate (CAGR) from 2021, reaching \$333.8 billion by 2027, taking into consideration the impact of the COVID-19 pandemic. The ecotourism market growth in these regions is the increasing motivation to travel to unique destinations, biodiversity, coral reefs, and pristine nature; the desire to explore undisturbed natural regions; and an increased focus on sustainability.[77]

Adventure Tourism

Adventure tourism involves traveling to a remote or exotic location to take part in physically challenging outdoor activities. They can be grouped according to the natural space in which the activity takes place—land, water, or air. According to the US-based Adventure Travel Trade Association, adventure travel may be any tourist activity, including two of the following three components: a physical activity, a cultural exchange or interaction, and engagement with nature. It is categorized as "hard" or "soft." The "soft" type implies those activities that involve less risk and physical demand. On the other hand, "hard" activities involve a certain level of risk and a certain skill.

[75]   R. C. Buckley. 2009. *Ecotourism: Principles and Practices*. Wallingford: CAB International; D. Fennell. 2003. *Ecotourism*. London: Routledge; and D. Weaver. 2008. *Ecotourism.*2nd edition. Brisbane: John Wiley and Sons.
[76]   R. C. Buckley. 2010. *Conservation Tourism*. Wallingford: CAB International.
[77]   Allied Market Research. 2021. Ecotourism Market: Global Opportunity, Analysis, and Industry Forecast, 2021-2027.

The market size of adventure tourism was estimated to be $112.23 billion in 2020, and to grow at a CAGR of 20.1% from 2021 to reach $1.169 trillion by 2028, taking into consideration the impact of the COVID-19 pandemic. The soft category, valued at $37.6 billion in 2020, is considered the most important and is expected to reach $380.69 billion by 2028, at a growth rate of 20.1% over the forecast period.

Cultural Tourism

Cultural tourism is defined by the UN World Tourism Organization as "A type of tourism activity in which the visitor's essential motivation is to learn, discover, experience, and consume the tangible and intangible cultural attractions/products in a tourism destination. These attractions/products relate to a set of distinctive material, intellectual, spiritual, and emotional features of a society that encompasses arts and architecture; historical and cultural heritage; culinary heritage; literature; music; creative industries and the living cultures with their lifestyles; value systems; beliefs; and traditions."[78]

At least 40% of tourism can be considered cultural tourism. There is a core market of tourists that travel primarily for culture and a vast majority that engage in cultural tourism when they travel as a secondary activity.[79]

Marine Tourism

Marine tourism refers to leisure activities that take place away from one's home and carried out in large bodies of water, such as the sea. The global marine tourism market was worth $67.44 billion in 2019, and it is predicted to grow to $106.72 billion by the end of 2026, with a CAGR of 6.7% between 2021 and 2026, taking into consideration the impact of the COVID-19 pandemic.

Nature-Based Tourism Innovations

In February 2022, the Kerala State Water Transport Department announced a new initiative as part of its commitment to sustainable methods specifically aligned to tourism activities. The new 24 m x 7 m double-decker vessel, Solar Cruiser, provides air-conditioned comfort for up to 200 backwater cruise passengers. It is fully solar-powered, with 100-kilowatt (kW) batteries.[80]

This type of investment in, and promotion of, sustainable travel methods within the tourism offer is also echoed in Spain, where visitors to Madrid are encouraged to join an electric bicycle (e-bike) tour to various cultural centers.[81] Tourists are informed that this will enable them to support urban regeneration, boost the local economy, and take part in responsible tourism.

Many tourism companies are increasingly keen to associate themselves with eco- and nature-based tourism activities. In the Caribbean, the Sandals Foundation has invested in many environmental education and advocacy programs: establishing marine sanctuaries, outplanting more than 12,000 corals, and engaging over 55,000 people in conservation efforts.[82]

78   World Tourism Organization. *General Assembly 22nd Session*.
79   World Tourism Organization. 2018. *Tourism and Culture Synergies*. Madrid: UNWTO.
80   J. L. Paul. 2022. SWTD to Launch First Solar-Powered Cruise Vessel from Kochi's Marine Drive in April. *The Hindu*. 18 February.
81   Authenticitys. Explore Madrid's Urban Regeneration on an E-Bike.
82   *Jamaica Observer*. 2022. Sandals Resorts Announces '40 for 40 Initiative' Projects. 18 March.

A number of organizations offer scuba diving expeditions that include marine biology training sufficient for participants to collect scientific data as part of the trips. Reef Check, Biosphere Expeditions, and Coral Caye are marine conservation NGOs that offer such trips.

Digital means are also being employed to educate tourists on sustainability matters and enhance their experience at the same time. For example, the Ol'au Palau initiative, through a gaming app, aims to promote responsible tourism that encourages tourists to complete eco-friendly measures.[83] A points-based system is adopted where users earn badges by completing tasks that include avoiding single-use plastics, visiting culturally significant tourism sites, and using sunscreen that is safe for the reefs. Travelers are also urged to take part in regenerative tourism projects, eat local and sustainably sourced foods, and support businesses that aim to reduce their environmental and cultural impact. Gamers can also score points by correctly answering questions about Palau's biodiversity and culture. The final set of rewards encourages tourists to offset their carbon footprint by using Palau's personal carbon calculator.[84]

Conclusion

Initial introductions to three sectors of interest to the MARES regenerative approach have been presented. Each of them, for varying reasons, offers great promise for future development, and each sector is populated with interesting innovations attracting investment.

However, while there are good articulations of need for each and encouraging innovations being developed, the scale of ambition within each of these sectors may not yet match with the potential size of the opportunity, and new forms of catalyzing are required. In the next section, the three areas of opportunity that jointly may bring about the required transformation will be identified.

---

83   Ol'au Palau.
84   Government of the Republic of Palau, Bureau of Tourism. Palau Offset Calculator.

2

POWERING THE FUTURE: THE PROMISE OF MARINE RENEWABLE ENERGY

CHAPTER 4

Harnessing Marine Renewable Energy

The first main blue economy opportunity is harnessing more marine renewable energy (MRE). The opportunity is significant, especially for large ocean states. Some estimates suggest that wave energy could power the entire world on its own,[85] but the pace of uptake still lags expectation. This chapter presents an overview of the main MRE platforms.

Marine Renewable Energy—The Blue Economy in Action

Within the mix of blue economy developments that aim to utilize, protect, and regenerate ocean resources, MRE has an important and crosscutting role. In fact, the greatest potential out of all the different renewable energy systems lies in the utilization of the energy within the seas and oceans. The inherent regenerative potential of MRE is becoming more promising as the underlying technologies progress. However, before moving to the specifics, it is worth considering the broader, crosscutting blue economy value that may be derived from MRE investments.

The Office of Energy Efficiency and Renewable Energy of the United States lists six key blue economy benefits of supporting renewable energy transition.[86]

- Improve access to resilient energy sources and clean water of remote, coastal, and island communities.
- Support the development of renewable energy-powered systems to observe, monitor, and protect the oceans.
- Encourage the development of marine energy-powered businesses and involve an emerging, diverse workforce in achieving ambitious renewable energy targets.
- Act as a supporting technology to increase the viability of hybrid systems and other renewable energy technologies.
- Utilize ocean energy to power new applications like ecotourism, sustainable aquaculture, and seawater mineral extraction.
- Promote technological development to support the decarbonization of the maritime sector.

Some of these opportunities will be explored in greater detail in this and subsequent chapters. But first an overview of some of the more prevalent MRE technologies will be explained in the succeeding sections.

---

85   World Economic Forum. 2022. *Wave Energy: Can Ocean Power Solve the Global Energy Crisis?* 22 March.
86   US Department of Energy, Office of Energy Efficiency and Renewable Energy. *Marine Energy and the Blue Economy.*

What Is Marine Renewable Energy?

MRE is defined by the European Science Foundation as "renewable energy production, which makes use of marine resources or marine space," such as offshore wind, marine floating solar photovoltaic (FPV), and ocean renewable energy (ORE).[87]

Some renewable energy technologies are placed in the ocean, but their energy production does not depend on ocean energy. Solar energy, for example, requires large surface areas to increase energy production. With available land costing high in many parts of the world, it is understandable why attention should turn to the deployment of solar energy panels within the vast area of the ocean by placing solar energy cells in floating carriages. This configuration is called floating solar.

Another renewable energy system that can utilize large ocean areas is offshore wind. This configuration places the wind turbines away from the shore into ocean areas solving the problem of space limitations on land for wind energy systems. Again, it does not rely directly on ocean energy, but is taking advantage of the large available area on the surface of the ocean.

As a subset of MRE, all forms of energy that can be derived directly from the resources of the seas and oceans can be collectively called ORE. These technologies include the conversion of currents (i.e., ocean, tidal in-stream); tidal range; wave; thermal gradient; and salinity gradient.

Further descriptions of the key MRE and ORE technology platforms follow, to provide context for the decision-making options that will be featured subsequently.

Types of Marine Renewable Energy Technologies and Approaches

Offshore Wind

Offshore wind energy is captured using large fan blades that resemble giant propellers located at the top of tall towers. The fan-like blades spin when the wind blows, and these spinning blades drive turbines that create electricity. The turbine towers can be deployed in the deep open ocean offshore or in near-shore shallow seas. Wind farms may also be installed in freshwater lakes, saltwater fjords, or coastal water bodies. The main advantage of offshore wind over its onshore counterpart is the higher and more consistent wind speeds. The downside is a more expensive overall cost because of higher deployment and installation costs, as well as all other balance-of-plant costs.

Two types of foundations are used to attach offshore wind towers to the seabed depending on the bathymetry (water depth) at a given location: (i) bottom fixed for depths of up to 70 m; and (ii) floating structures for depths of up to 1,000 m. In general, the deeper the location, the more expensive the technology and deployment costs. Furthermore, deeper areas usually mean greater distances from the shore, which increases the costs due to the need for longer subsea power cables to bring the electricity onshore. However, the advantages of more distant offshore sites are the greater areas of potential use as wind farms and lesser intrusion on human activities, thus providing added benefits to the overall acceptability of the wind farm.

---

[87]   European Science Foundation. 2010. *Marine Renewable Energy: Research Challenges and Opportunities for a New Energy Era in Europe*. Marine Board Vision Document 2.

Figure 1 presents a global map of the wind speeds at 100 m height. Wind speeds of at least 7 meters per second (m/s) are viable for offshore wind projects and are shown in dark orange to red. This map clearly shows that wind speeds are high enough near the coast of many Asian countries and the Americas. However, wind speed is only the first level of consideration in determining the viability of a site for a wind farm. Assessing the technical potential of a site requires knowing not only wind speed, direction, and hours per day, but also bathymetry. In addition, some locations may not be suitable due to proximity to endangered bird flyways, navigation channels, or onshore and offshore structures, e.g., oil rigs. The economic viability of a wind farm also depends on the cost of electricity in each area. The best location for a wind farm is determined after considering all other constraints such as social aspects of fairways and navigational pathways, environmental factors such as migration routes, and logistic factors pertaining to construction and operation.

In 2019, the total technical potential for offshore wind in select countries with emerging markets was estimated at 3.082 terawatts (TW).[88] The countries studied include Brazil, India, Morocco, Philippines, South Africa, Sri Lanka, Turkey, and Viet Nam. Of the total estimated potential, bottom fixed type is approximately 1.016 TW, while floating is at 2.066 TW. Globally, the technical potential of offshore wind is said to be around 36,000 terawatt-hour (TWh) per year for bottom fixed installations within a 60 km distance from the shore, while the current demand is 23,000 TWh.[89] If deeper and farther locations are considered, the total potential is seen to be 11 times over the projected demand. By 2018, the total installed capacity of offshore wind was estimated to be 23 gigawatts (GW), which includes 4.3 GW added that year. The UK leads in installed capacity of 8 GW, with Germany coming in second with 6.5 GW and the PRC third with 3.6 GW.

Figure 1: Global Overview of Onshore and Offshore Wind Speeds within 200 Kilometers from Shore at 100 Meters Elevation

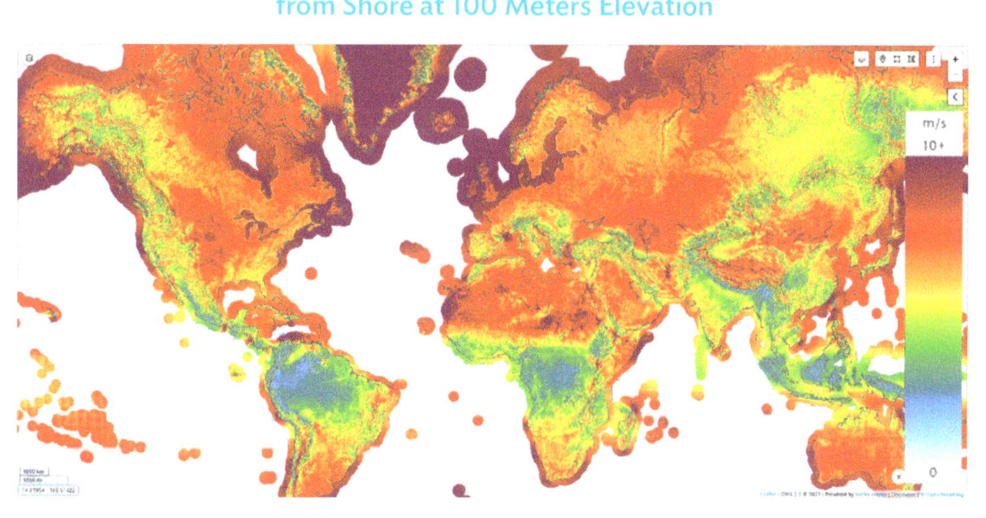

m/s = meters per second.

Source: Map obtained from the Global Wind Atlas 3.0, a free, web-based application developed, owned and operated by the Technical University of Denmark (DTU). The Global Wind Atlas 3.0 is released in partnership with the World Bank Group, utilizing data provided by Vortex, using funding provided by the Energy Sector Management Assistance Program (ESMAP). For additional information: https://globalwindatlas.info.

---

88 Energy Sector Management Assistance Program (ESMAP). 2019. *Going Global: Expanding Offshore Wind to Emerging Markets.* Washington, DC: World Bank.
89 International Energy Agency. 2019. *Offshore Wind Outlook 2019.* Paris.

Offshore wind farm costs have fallen by as much as 30% from 2001 to 2015. As a result, the levelized cost of energy (LCOE) has dropped from $240/megawatt-hour (MWh) in 2001 to $170/MWh in 2015.[90] More than half of the cost reduction is due to lower turbine costs. Improvements in the blade and drivetrain technologies will produce larger turbines with higher capacities that will help drive the cost of generation further down. Other areas of potential reductions in costs are improved foundation technologies for floating structures and better electrical interconnection strategies and technologies. They promise to bring costs down, making offshore wind more competitive than fossil fuel technologies and even onshore wind. By 2018, the LCOE had dropped to $140/MWh and was seen to go as low as $100/MWh when weighted average cost of capital is considered to reduce from 8% to 4% (footnote 89).

Floating Solar Photovoltaics

FPV mounts solar panels on a floating structure typically located on a calm body of water such as a reservoir, lake, or bay that is protected from waves.

FPV is among the world's most prevalent forms of renewable energy system. It is found in both freshwater and marine environments. It is especially attractive for countries and areas where availability of space on land is limited or expensive. An audit of the 10 largest global projects for FPV demonstrates that the largest projects are found in Asia, where there are high population densities and land values (Table 1).

Table 1: Global Floating Solar Farms
(updated 26 July 2021)

Capacity	Location	Details
2.2 GW	Indonesia	Badan Pengusahaan Batam to supply Singapore
2.1 GW	Republic of Korea	Saemangeum floating solar energy project
600 MW	India	Omkareshwar Dam floating solar farm
320 MW	PRC	Hangzhou Fengling Electricity Science Technology's solar farm
150 MW	PRC	Three Gorges New Energy's floating solar farm
145 MW	Indonesia	Cirata Reservoir floating photovoltaic power project
105 MW	India	NTPC Kayamkulam solar project
100 MW	India	NTPC Ramagundam solar power plant
70 MW	PRC	CECEP's floating solar project
60 MW	Singapore	Sembcorp's Tuas floating solar project

GW = gigawatt, MW = megawatt, PRC = People's Republic of China.
Source: Power Technology. 2021. Top Ten Operating Floating Solar Farms by Capacity.

The components of a marine FPV system include a floating platform; a system for mooring the platform; the array of solar photovoltaic (PV) modules; the cables needed to transfer the alternating current or direct current power generated; and the connectors for all the components.

The largest example of FPV will be installed in Indonesian waters (Figure 2). The project is expected to cover 1,600 hectares and will be combined with more than 4,000 MWh of battery storage. The $2 billion project will be located 20 km away from Singapore, across the Singapore Strait, and will supply Singapore via an undersea cable.

---

[90]   International Renewable Energy Agency (IRENA). 2016. *Innovation Outlook: Offshore Wind*. Abu Dhabi.

Figure 2: 2.2-Gigawatt Floating Solar to Supply Singapore from Indonesia

Source: Renew Economy (2021). https://reneweconomy.com.au/.

Tidal Energy

Tides are created by the moon's gravitational pull as it circles the earth and modified by the sun's gravitational pull. Tidal currents are created as the tides rise and fall, particularly when the water is forced to flow into shallow water or a narrow channel. Tidal currents can be harnessed in two different ways.

A dam-like structure can be installed in a selected body of water (usually near the shore) to allow a difference in water level on the two sides, e.g., as the tide falls. The potential energy created by the difference in water level can be harnessed by allowing the water to flow through turbines located within the structure. This technology is called tidal range or barrage.

Another system is based on in-stream tidal turbines, which use the kinetic energy of the tidal currents created by the ebb and flow of tides. Typical flow speed for tidal turbines is at least 1.5–2.0 m/s.

Tidal ranges vary greatly with location, from a few meters to a few centimeters, as can be seen in the global tidal range distribution chart (Figure 3), making the technology site-dependent. Because of this constraint, it has the smallest potential of any ocean renewable energy, with an estimated energy of 1,200 TWh per year.[91]

---

[91] IRENA. 2014. *Tidal Energy: Technology Brief.* Abu Dhabi.

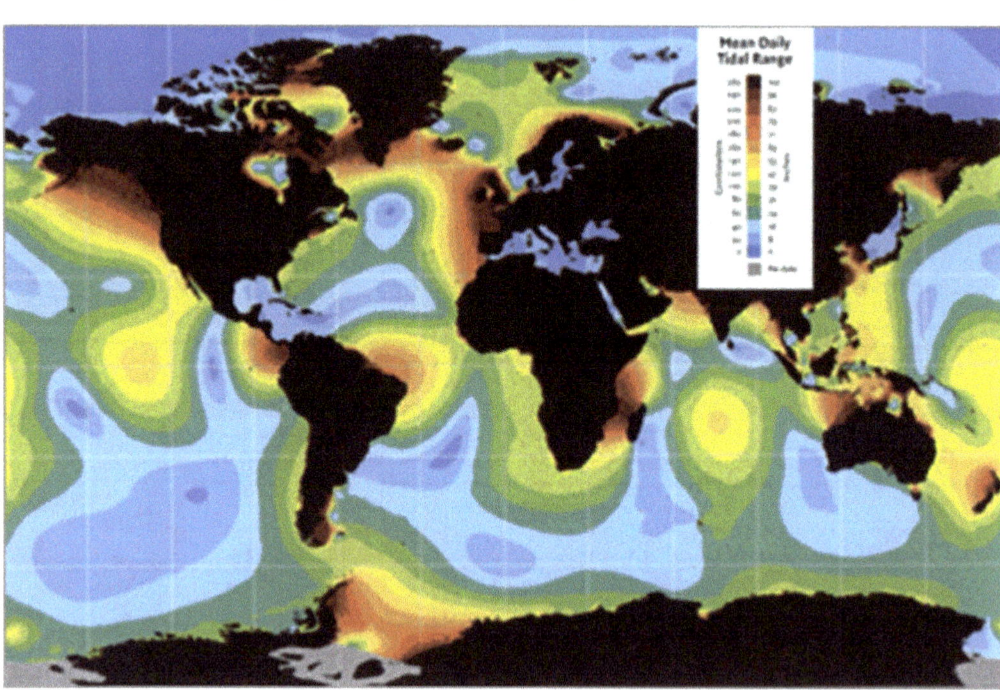

Figure 3: Global Tidal Range Distribution

Source: Renew Economy (2021). https://reneweconomy.com.au/.

Tidal Range and Barrages

Tidal range technology utilizes the potential energy created by the difference in water level between the two sides of a dam-like structure called a barrage (Figure 4a). During high tide, water can be trapped on one side of the dam and as the tide ebbs on the other side, gravity will cause the trapped water to flow down and out of the basin. This water can pass through a channel where a turbine is in a small tunnel at the base of the dam. As the turbine turns, it will extract energy from the water flowing out. To allow energy extraction from both the entry and exit of water, two-way turbines are usually installed in tidal barrages.

Tidal barrage has a technology readiness level (TRL) of 9 and has the largest share (98%) of installed MRE in the world.[92] The La Rance tidal power plant in France (Figure 4b) is the first tidal barrage system to be fully operational. It is a 240-megawatt (MW) power plant and is currently the second largest tidal barrage in the world—the Sihwa Lake plant in the Republic of Korea is the largest, with a capacity of 254 MW (footnote 91).

---

[92]   IRENA. 2020. *Innovation Outlook: Ocean Energy Technologies.* Abu Dhabi.

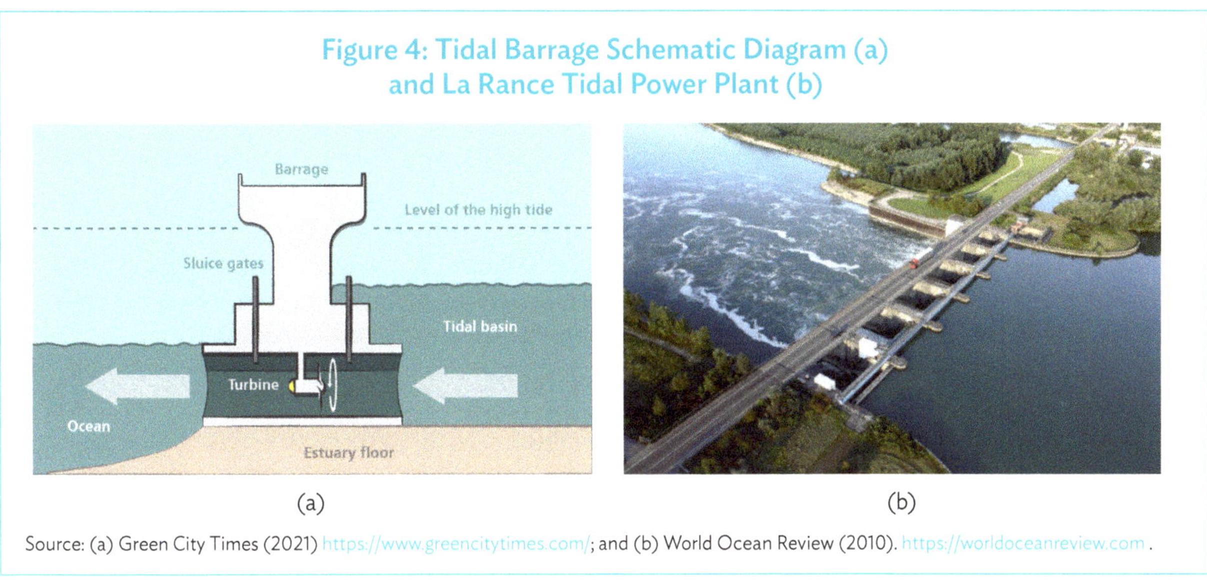

**Figure 4: Tidal Barrage Schematic Diagram (a)
and La Rance Tidal Power Plant (b)**

Source: (a) Green City Times (2021) https://www.greencitytimes.com/; and (b) World Ocean Review (2010). https://worldoceanreview.com .

Tidal Currents (Tidal Stream)

In-stream tidal turbine technology extracts power directly from the kinetic energy of tidal currents (incoming and outcoming flow due to tides), with a flow velocity of at least 1.5–2.0 m/s. It is currently rated as a TRL 8 and is projected to become more common than tidal barrages in the near future because it does not require construction of a barrage structure (footnote 92). The most common device used for this technology is the horizontal-axis tidal turbine (HATT), shown in Figure 5. The HATT functions like a horizontal-axis wind turbine and has a current capacity range of 100 kW to 1.5 MW. Other in-stream devices being developed are the venturi-type HATT and the vertical-axis tidal turbine.

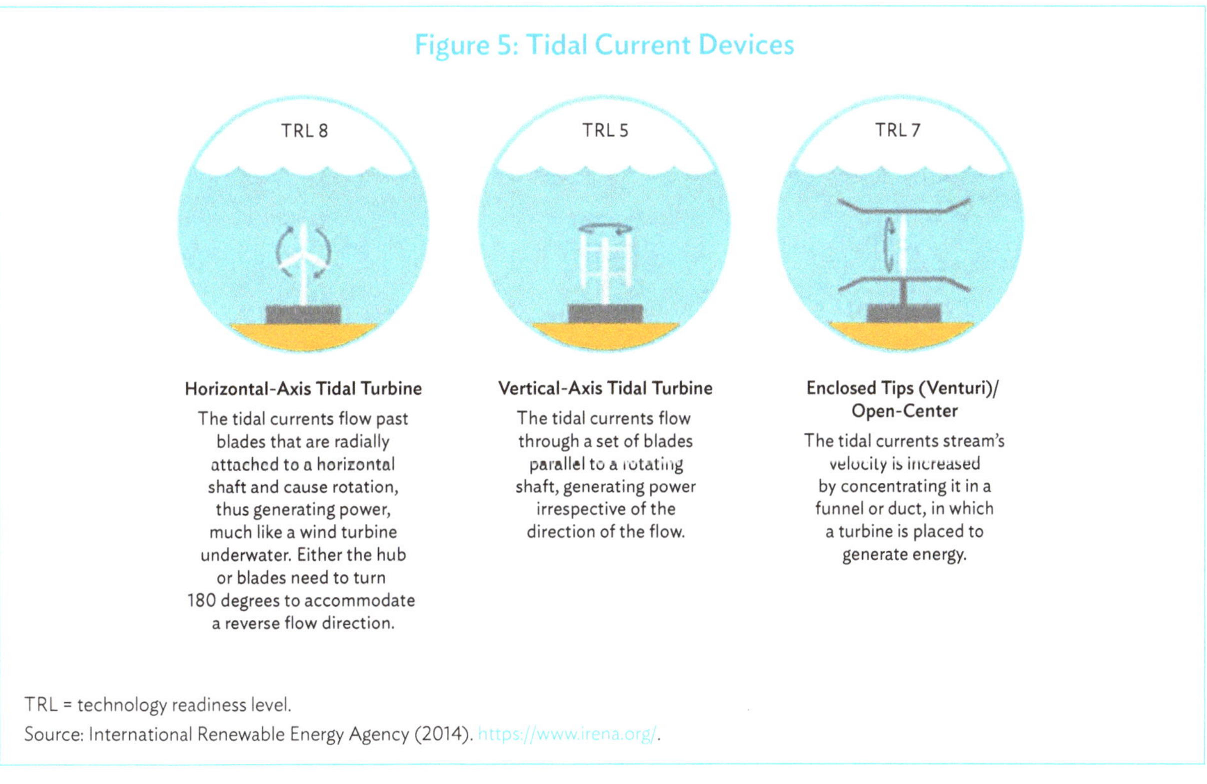

Figure 5: Tidal Current Devices

Horizontal-Axis Tidal Turbine

The tidal currents flow past blades that are radially attached to a horizontal shaft and cause rotation, thus generating power, much like a wind turbine underwater. Either the hub or blades need to turn 180 degrees to accommodate a reverse flow direction.

Vertical-Axis Tidal Turbine

The tidal currents flow through a set of blades parallel to a rotating shaft, generating power irrespective of the direction of the flow.

Enclosed Tips (Venturi)/ Open-Center

The tidal currents stream's velocity is increased by concentrating it in a funnel or duct, in which a turbine is placed to generate energy.

TRL = technology readiness level.
Source: International Renewable Energy Agency (2014). https://www.irena.org/.

Tidal stream turbine systems have a current installed capacity of around 10.6 MW but are predicted to increase up to 570 MW once the planned multidevice tidal arrays are installed. One of the major tidal stream installations currently being developed is the Meygen Ltd project, a 398 MW project that is currently in the Deployment Phase 1 (footnote 92).

Wave

Waves in the sea are created by wind blowing over the surface of the water. The higher the wind speed, the longer time the wind blows over the same area, and the larger the distance (fetch) the wind blows over, then the higher the waves created. Once created by passing storms, large waves can travel thousands of kilometers across deep ocean waters without losing much energy. Wave energy is the transfer and conversion of energy from ocean waves. The energy captured is utilized for different kinds of useful work, such as electricity generation, water desalination, and water pumping. Wave energy is estimated to be the largest global form of ocean energy.[93] Since solar energy drives the differences in air temperature that cause wind, wave energy is considered a form of concentrated solar energy.

The largest waves are created by powerful winds located between 30 degrees and 60 degrees latitude in both the northern and southern hemispheres (Figure 6), with the largest power levels occurring off the west coast of continents and in southern waters.[94] The movement of waves depends on the wave height, wave speed, wavelength (or frequency), and water density. Compared to tidal, wave energy resources are more spatially distributed. Theoretically, the global annual potential of wave energy is 500 TWh, which indicates that it could meet the entire global energy demand (footnote 92). Although waves vary seasonally and in the short term, their behavior can be predicted using models and monitored using buoys and satellites. They are considered a reliable energy source.

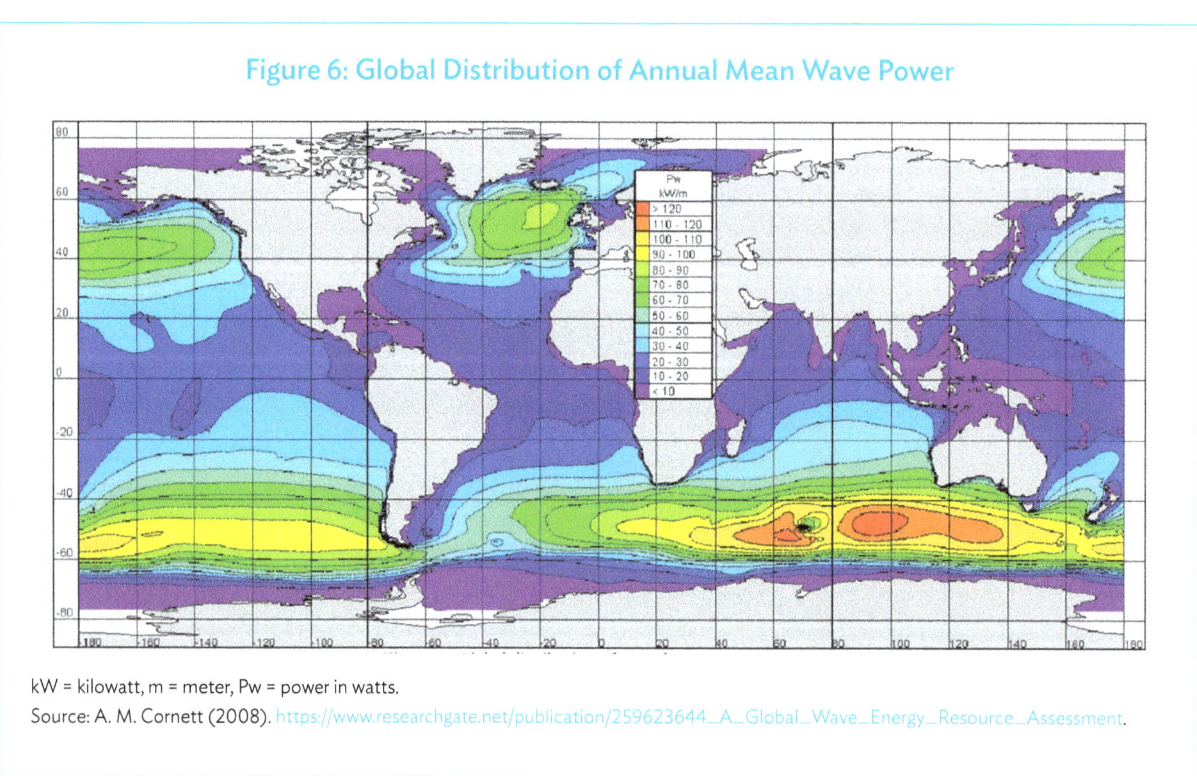

Figure 6: Global Distribution of Annual Mean Wave Power

kW = kilowatt, m = meter, Pw = power in watts.
Source: A. M. Cornett (2008). https://www.researchgate.net/publication/259623644_A_Global_Wave_Energy_Resource_Assessment.

---

[93]   A. Clément et al. 2002. Wave Energy in Europe: Current Status and Perspectives. *Renewable and Sustainable Energy Reviews.* 6 (5). pp. 405–431.
[94]   IRENA. 2014. *Ocean Energy: Technology Readiness, Patents, Deployment Status and Outlook.* Abu Dhabi.

Wave energy conversion is carried out by transforming either or both kinetic and potential energy of ocean surface waves into a more desired form such as electricity. Wave energy converters can absorb kinetic and potential energy through moving bodies and attenuators, respectively, while both energies can be absorbed through point absorbers. As has been observed for wind energy, wave energy technologies have not seen a convergence toward one type of design. The three main working principles that have emerged include the oscillating water columns to compress air to drive an air turbine, oscillating bodies converters that use a variety of devices to transform wave motion between bodies, and overtopping devices that use the potential energy of water spilling into a closed reservoir to drive a hydraulic turbine.[95] Despite the high potential for wave energy, it has lagged other forms of renewable energy, perhaps because of the lack of a standardized design such as wind turbines.

Ocean Thermal Energy Conversion

Ocean thermal energy conversion (OTEC) takes advantage of the temperature difference between the warm upper surface of the oceans and the cold deep layers. A heat engine can be used to utilize the temperature gradient but needs a 20°C difference for a conversion cycle to work (usually Rankine cycle). At 800 m to 1,000 m deep below the ocean surface, the temperature is quasi-constant at around 4°C, suggesting that the surface of the ocean should be averaging 25°C for the system to work. This constraint limits the technology, making it dependent on sites with waters of 1,000 m depth or more in between 30° north latitude and 30° south latitude (Figure 7) (footnote 92).

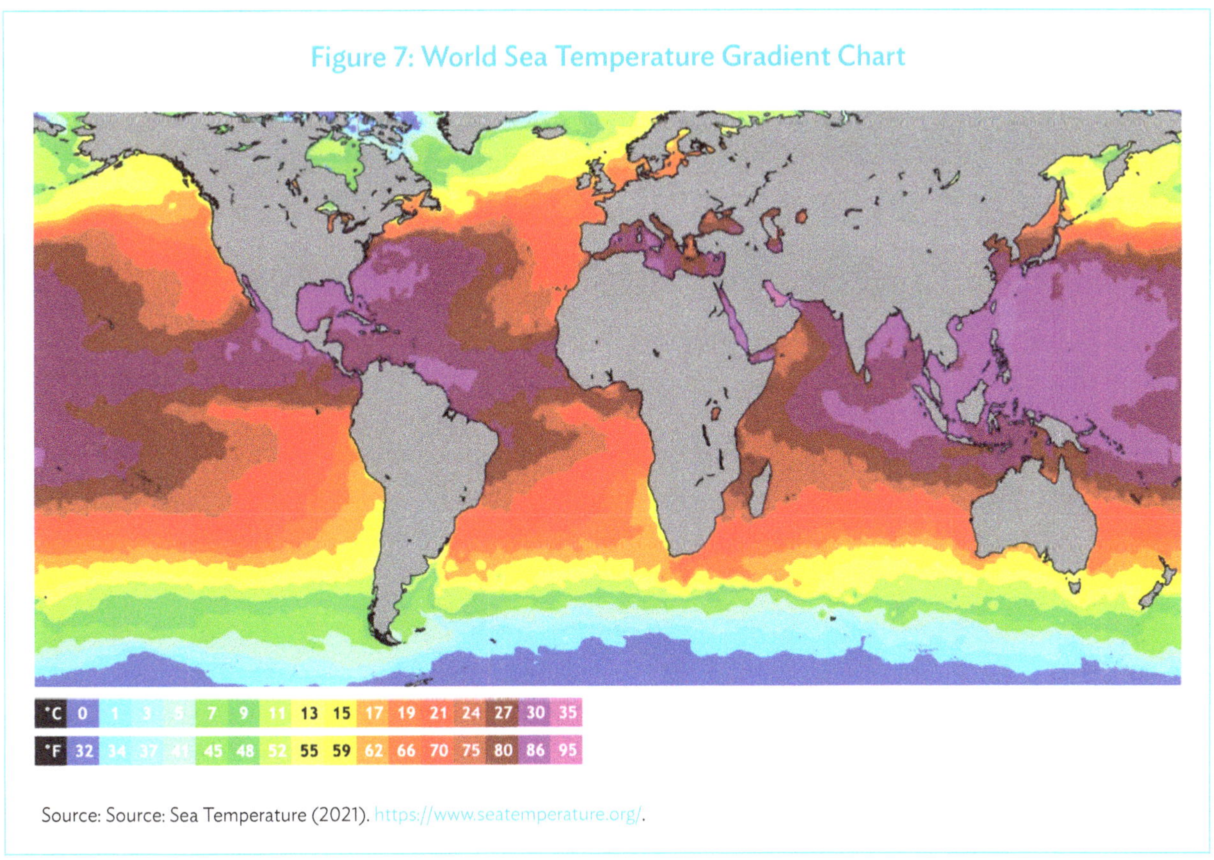

Figure 7: World Sea Temperature Gradient Chart

Source: Source: Sea Temperature (2021). https://www.seatemperature.org/.

There are four main types of OTEC:

- Open-Cycle OTEC. This uses water as the main working fluid. Warm surface water is vaporized by allowing it to flow through a valve into a low-pressure containment. The vapor is used to drive a turbine and then the spent water is condensed using the cold water from the lower layers of the ocean. This technology allows power production, freshwater production, and possible air-conditioning from the pumped cold waters.

- Closed-Cycle OTEC. This is the well-known thermal Rankine cycle where a low boiling point working fluid (ammonia, refrigerants) is used to drive a generator. Closed-cycle is more efficient when compared to open-cycle OTEC.

- Kalina Cycle OTEC. A variation of the closed-cycle, further increasing the efficiency by using water-ammonia mixture as working fluid.

- Hybrid OTEC. This system takes advantage of both the open- and closed-cycle OTEC. The power generation is the same as the closed-cycle OTEC, but the spent seawater from the heat exchangers is flash evaporated and condensed using cold water to produce fresh water (footnote 92).

The OTEC Foundation has confirmed the significant potential for OTEC globally, at least for a 30-MW plant facility or equivalent combination of smaller plants (Figure 8). The Organization for the Promotion of Ocean Thermal Energy Conversion and GEC Co. Ltd. of Japan proposed an OTEC plant for Kwajalein Atoll.[96]

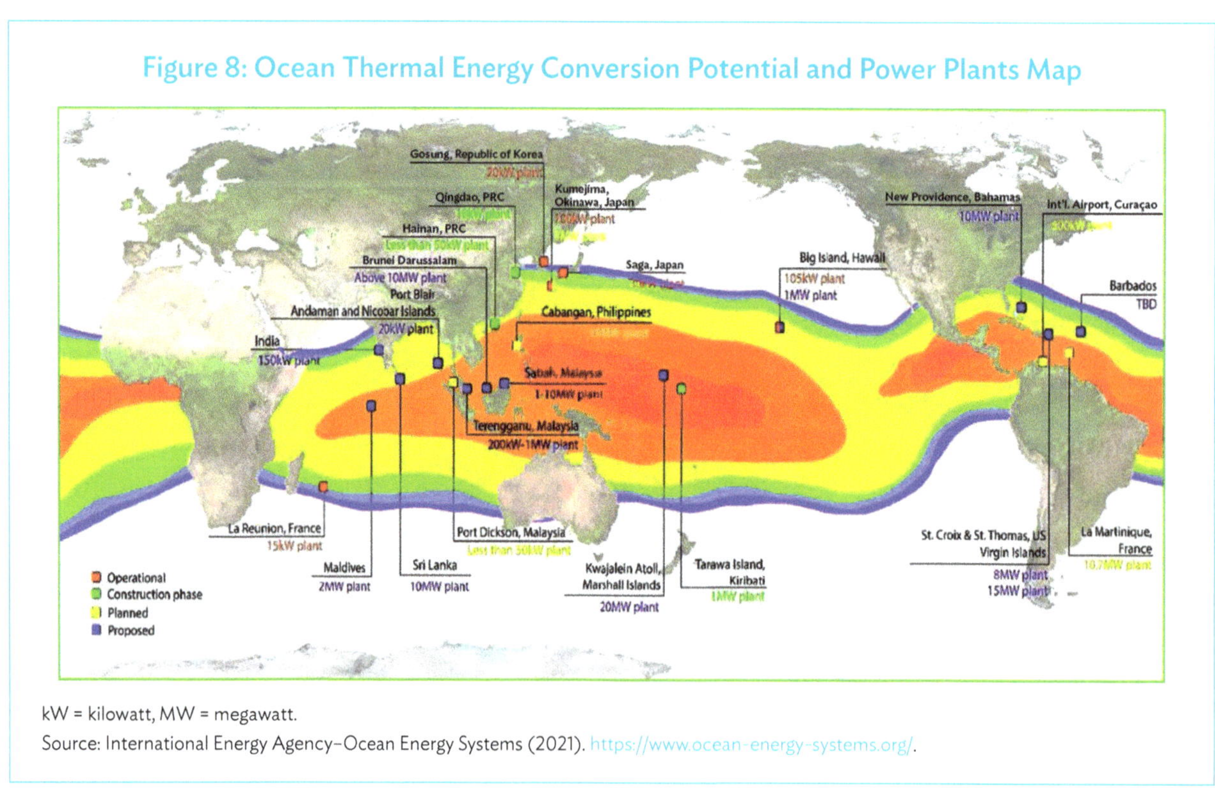

kW = kilowatt, MW = megawatt.
Source: International Energy Agency–Ocean Energy Systems (2021). https://www.ocean-energy-systems.org/.

---

[96] IRENA. 2013. *Pacific Lighthouses: Renewable Energy Opportunities and Challenges in the Pacific Islands Region—The Republic of the Marshall Islands.* Abu Dhabi.

The technology is currently in the research and development phase for scaling up. The largest working OTEC plant is the 1 MW plant in Hawaii, but it was decommissioned in the late 1990s. There are various 100 kW capacity plants in Japan and the Republic of Korea, but they are reserved mostly for research and demonstration purposes. Current projects in the developmental stage are the 10 MW OTEC plants being developed in Hawaii, south of the PRC and France (footnote 92).

Salinity Gradient

There are two different types of salinity gradient power generation systems currently being investigated. The first is known as the reverse electrodialysis (RED) system. The second and more recent approach is known as pressure-retarded osmosis (PRO). Both are still in the research phase, with costs and power efficiencies still under evaluation.

The RED system uses a stacked approach, wherein alternating cation and anion membranes are placed between a cathode and anode in alternating patterns (Figure 9). Typically, there is a stack of three cells. The chemical potential difference enables the movement of ions through the membranes from the concentrated solution to the diluted solution. Sodium chloride solution, sodium ions permeate through the cation exchange membrane in the direction of the cathode and chloride the other way around.

The system uses both seawater, in a concentrated solution, as well as river water, in a diluted solution, to cause the chemical reaction. The process is clean, with the only effluent being brackish water.

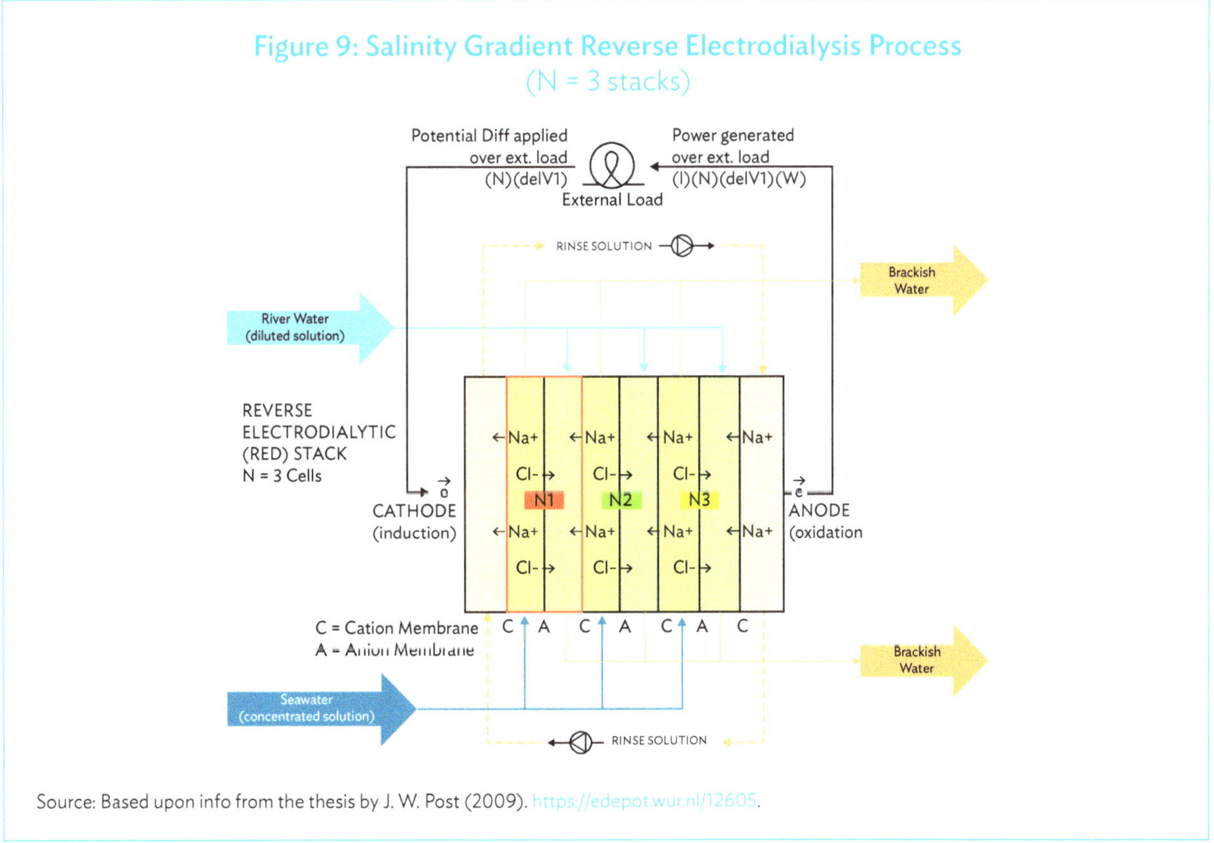

Figure 9: Salinity Gradient Reverse Electrodialysis Process (N = 3 stacks)

Source: Based upon info from the thesis by J. W. Post (2009). https://edepot.wur.nl/12605.

The PRO system is quite different and holds promise to be even more cost-effective (on a generated power $/kilowatt-hour [kWh] basis) than the original RED approach, even though the system includes mechanical systems (turbine or generator, pressure exchange vessel). The central pressure vessel uses permeable membranes, separating the high-pressure seawater (in a concentrated solution) from the river water (in a diluted solution). The membranes allow the seawater to permeate while retaining the solute (i.e., dissolved salts). A higher-pressure solution is created when river water from the low-pressure diluted solution is transported to the high-pressure concentrated solution. The pressurization of the transported water is used to drive a turbine, which is connected to a generator through a common shaft. The rotating generator produces electrical power. Once again, the production of power through the PRO system is considered extremely clean, since the only effluent is the brackish water expelled from the system (Figure 10).

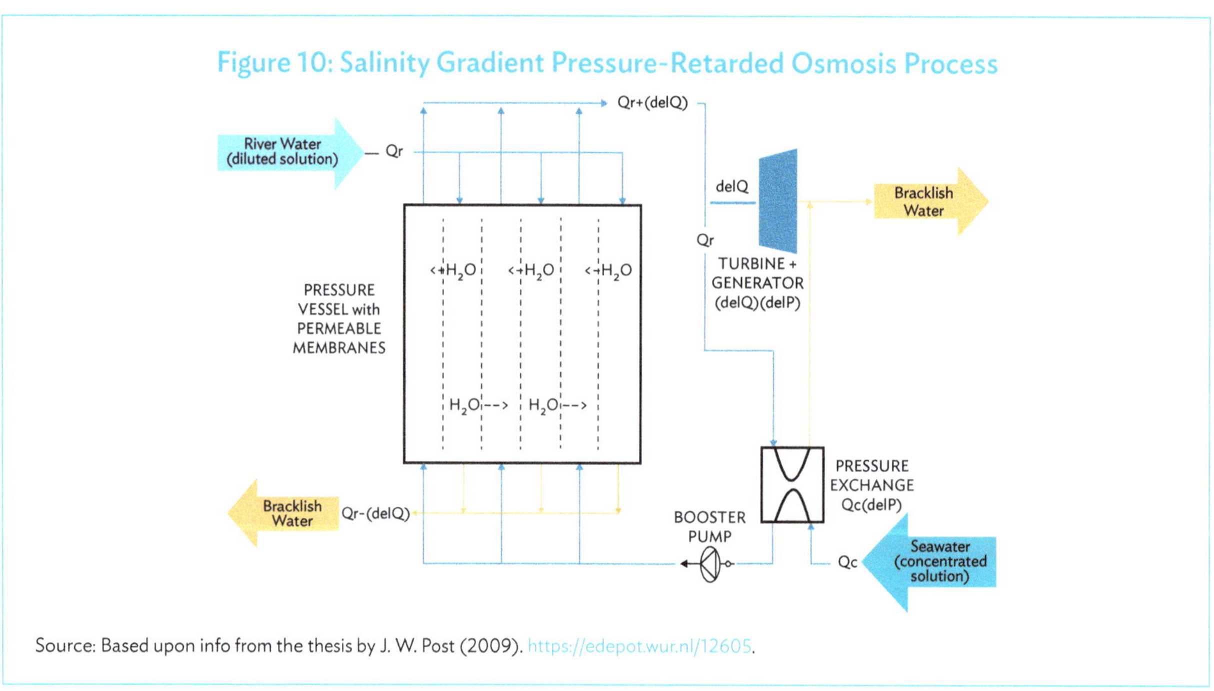

Figure 10: Salinity Gradient Pressure-Retarded Osmosis Process

Source: Based upon info from the thesis by J. W. Post (2009). https://edepot.wur.nl/12605.

Research has shown a strong potential for economic advantage from the use of salinity gradient power. Table 2 shows calculated potential over other energy sources. The table also demonstrates that although the price for generated electricity using the RED system is still high, the PRO system has potential to be competitive with other renewable energies.

Table 2: Comparison of Traditional and Renewable Energies with Salinity Gradient

Energy Source	GHG Emissions (g CO$_2$-e/kWh)	Price of Generated Electricity ($/kWh)	EROI	Availability of Renewable Sources	Efficiency of Energy Conversion (%)
Photovoltaic	90	0.24	1.6–6.8	Dependent	4–22
Wind	25	0.07	18	Dependent	24–54
Hydro	41	0.05	>100	Always available	>90
Geothermal	170	0.07	n/a	Dependent	10–20
Coal	1,004	0.042	80	Nonrenewable	32–45
Gas	543	0.048	10	Nonrenewable	45–53
SGP RED	<10	0.10	7	Always available, but not in noncoastal countries	50–70
SGP (PRO)	<10	0.065–0.130 with subsidies $0.05–$0.06	6–7	Always available, but not in noncoastal countries	40

CO$_2$-e = carbon dioxide equivalent, EROI = energy return on invested, g = gram, GHG = greenhouse gas, kWh = kilowatt-hour, n/a = not applicable, PRO = pressure-retarded osmosis, RED = reversed electrodialysis, SGP = salinity gradient power.
Source: A. Mora and A. de Rijck. 2015. Blue Energy: Salinity Gradient Power in Practice. GSDR Brief. Global Sustainable Development Report. United Nations Sustainable

Deep Ocean Current

Deep ocean currents hold enormous potential, and much of that potential remains unexplored. Ocean currents have the advantage over tidal currents in that they are relatively continuous, versus the periodic nature of tidal currents. Currents are driven by a variety of factors, including temperature, surface wind, the topography of the ocean floor, variations in water salinity, and the rotation of the earth. The most important factors driving ocean currents are heating of the water as well as wind effects. The variations in density and salinity also have an important effect on ocean currents. As Figure 11 shows, currents are found throughout the world, and thus are an important opportunity for potential power generation.

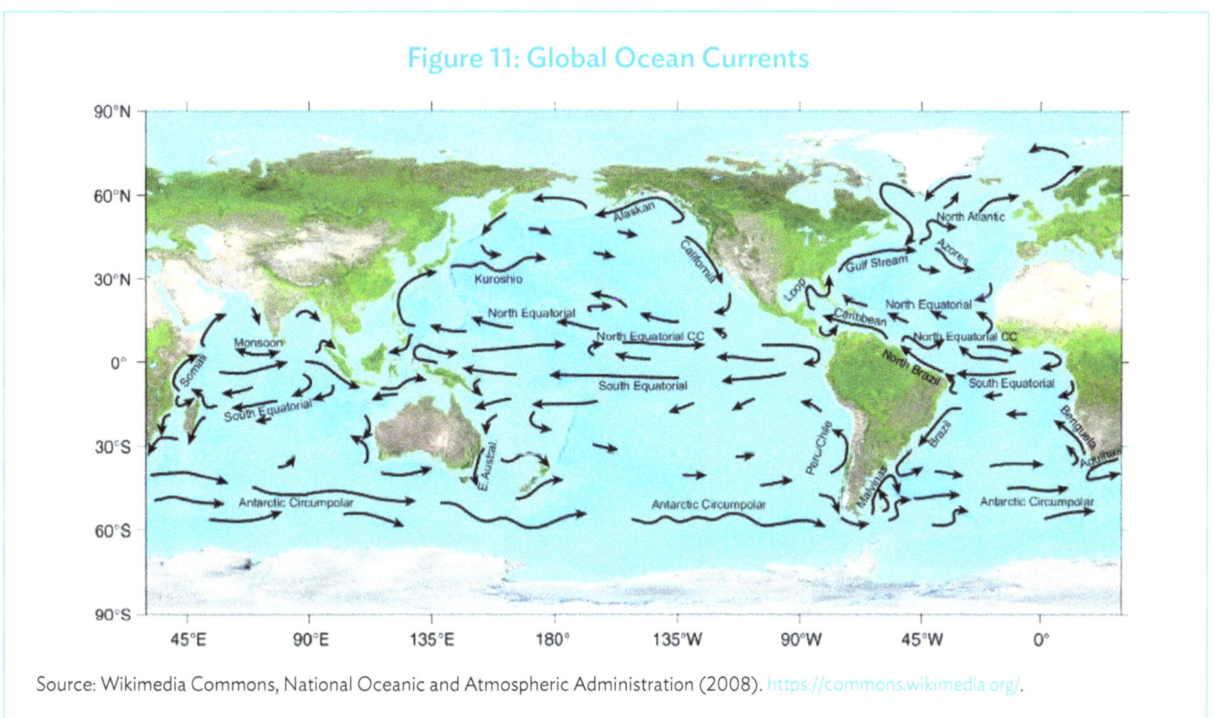

Figure 11: Global Ocean Currents

Source: Wikimedia Commons, National Oceanic and Atmospheric Administration (2008). https://commons.wikimedia.org/.

Understanding of the dynamics of ocean currents, which is fundamental to advancing the state-of-the-art in deep ocean current energy development, is computationally intensive and being improved. Other technologies are being employed, such as satellite tracking of autonomous buoys. These are useful for understanding surface currents and rates, but do not provide much data for the underlying currents beneath the surface. For this, complex computation models are employed. For example, Georgia Tech University in the US has a dedicated Ocean Science and Engineering Department. Their earth systems modeling and data sciences section have made important contributions to understanding the potential for deep ocean current power generation. Their ability to model deep ocean currents is a fundamental enabler for the initial understanding and prefeasibility development of profiles for proposed energy generation sites.

The technology readiness for deep ocean current is still very young. In 2017, IHI Corporation was able to test the world's first 100 kW ocean turbine off the coast of Japan (Figure 12).[97] The turbine was installed in the Kuroshio current, off the coast of Kuchinoshima Island, Kagoshima Prefecture. Calculations from the data obtained an estimate that the potential of the current amounts to 205 GW, demonstrating the promising potential of deep ocean current technology.

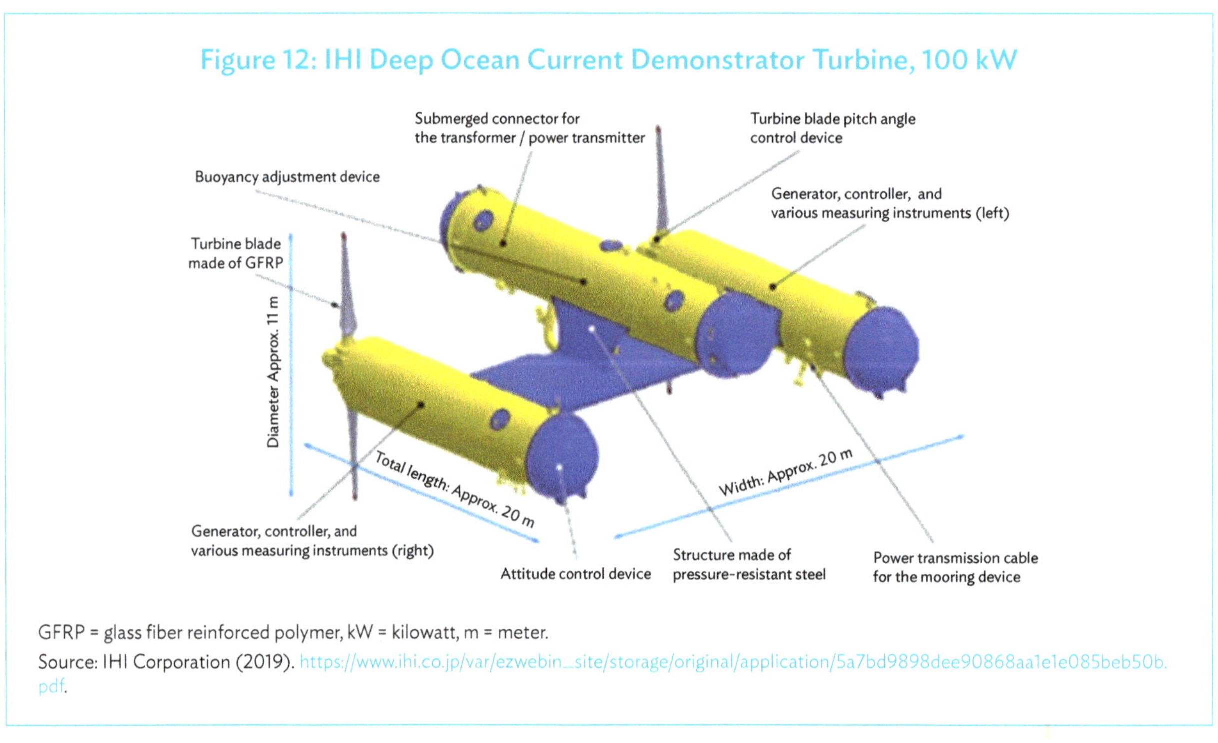

Figure 12: IHI Deep Ocean Current Demonstrator Turbine, 100 kW

GFRP = glass fiber reinforced polymer, kW = kilowatt, m = meter.
Source: IHI Corporation (2019). https://www.ihi.co.jp/var/ezwebin_site/storage/original/application/5a7bd9898dee90868aa1e1e085beb50b.pdf.

The potential environmental impacts of deep ocean current energy are another important area of research. A recent study discovered journals, with 561 articles tagged with "Marine Renewable Energy" and 752 articles tagged with "Ocean Energy." This included a total of over 16,000 articles identified in the field. Clearly there is a great deal of interest and research being shown in this field.[98] However, despite this plethora of research, many of the technologies involved in deep ocean currents are still new. Thus, all current studies regarding potential impacts in the environment necessarily focus on prototypes and demonstration devices.

---

[97]   IHI Corporation. 2019. *IHI Demonstrated the World's Largest Ocean Current Turbine for the First Time in the World*. *IHI Engineering Review*. 52 (1). pp. 6–9.
[98]   M. Martinez et al. 2021. *A Systemic View of Potential Environmental Impacts of Ocean Energy Production*. *Renewable and Sustainable Energy Reviews*. 149. 111332.

As the technology matures and more substantial efforts are implemented, the impact upon the environment can be further studied in more detail. These studies are extremely important since the current literature provides very little substantive information. In fact, considering the extensive review performed in the study by Martinez et al. (2021) on the potential environmental impacts of ocean–energy production (footnote 98), the magnitude of the environmental impacts is seldom mentioned in the studies. Thus, research on the environmental impacts of deep ocean current technologies is still to be quantified, along with mitigation strategies to ensure a safe implementation. This is yet another indication of the infancy of this technology.

Marine Bioenergy (Experimental)

Bioenergy refers to organic-derived electrical energy being produced, such as algae (macro and micro), cyanobacteria, regular bacteria, seaweed, kelp, algae, biomass, and other compounds. Production processes involve harvesting, extraction, pyrolysis, gasification, liquefaction, processing, and anaerobic digestion. Examples have been restricted to experimental pilot project form, including algal biofuels and cross-sector interlinkages with marine biotechnology.[99]

One marine bioenergy application is for shipping biofuels to ensure greener, less emissions-intensive transport. This sector is projected to accelerate in value because of the mandatory 2020 International Maritime Organization Sulphur Cap fuel emissions regulations and a 50% reduction target of total global emissions by 2050. Advanced biofuels have yet to fully commercialize beyond the experimental and/or piloting phase, especially in relation to MRE.[100] The global biofuel market for transport in general hovers at only around 1% of the total, given stakeholder uncertainty over potential rates of return on investment, consumer, and regulator responses. One experiment case study in the Republic of Korea recognized the possibilities in microalgae, but costs exceed $10 per kilogram of inputs to produce.[101] It identified the opportunities offered by permeable membrane technology and seawater nutrients.

Terrestrial-based bioenergy faces land constraints and barriers in securing reliable access to sufficient inputs to be upscaled, while marine sources such as kelp, algae, seaweed, etc. remain highly underdeveloped and proven economically and from a research perspective.

While marine bioenergy offers a perpetually renewable resource with few costs, when growing macroalgae in the vicinity of coral reefs, the organic compounds and microbial communities released by the seaweed have a negative impact on corals. Spore production by the seaweed may also lead to a further imbalance in macroalgal and/or coral competition. This has been well outlined by studies in the Line Islands and in other localities where a combination of human-induced eutrophication and algal growth leads to degradation of reef ecosystems.

[99]   US Department of Energy – Energy Transitions Initiative. Marshall Islands Energy Snapshot.
[100]   IRENA. 2019. *Future of Wind: Deployment, Investment, Technology, Grid Integration and Socio-Economic Aspects.* Abu Dhabi.
[101]   G. Hochman and C. Tabakis. 2020. Biofuels and Their Potential in South Korea. *Sustainability.* 12 (17). 7215.

Comparisons and Maturity

The various technological elements that make up the MRE sector all have unique challenges and benefits and are all at various stages of development. Table 3 provides a high-level view of the differences of eight key approaches.

Table 3: Summary of Marine Renewable Energy Technologies and Their Status

Technology	Techno-Economic Status	Challenges	Benefits
Solar PV (for Marine/ Offshore)	Mature Technology, Commercially Viable	Preferably deployed in calm waters; need proper planning for sea-space usage; address potential environmental concerns	Potential increased efficiency (due to cooling effect of water environment), co-application (e.g., food, water, transport, etc.)
Offshore Wind	Mature Technology, Available, Commercially Viable	Need for large marine area use an issue for large-scale farms	Leverage local marine and offshore capabilities (e.g., ship building, etc.)
Tidal Range (e.g., Barrage)	Mature Technology, Commercially Viable	Very high environmental impact for dam-like structures; Tide range >4 meters needed to be economically viable	Very predictable; technology similar to hydropower
Tidal Stream	Pre-Commercial, Turbine Technology High TRL	Technology costs need to be reduced	Very high predictability (18.6 years); resource not affected by weather; suitable for island regions
Ocean Current	Prototype and Pilot Systems	Limited sites; can be remote offshore	Can be baseload
Wave	Prototype and Pilot Systems	No technology convergence to scale (yet)	Niche applications; can be coupled to hybridize wind projects
OTEC	Prototype and Pilot Systems	Still very high LCOE (high CAPEX) even at 10 MW scale	Can be baseload supply
Salinity Gradient	Prototype and Pilot Systems	Still high LCOE. Needs co-application/integration (e.g., energy recovery in desalination plants) to make economic sense	Synergistic niche applications (aquaculture, salt production, desalination) and areas may be pathway to commercialization

CAPEX = capital expenditure, LCOE = levelized cost of energy, MW = megawatt, OTEC = ocean thermal energy conversion, PV = photovoltaic, TRL = technology readiness level.

Sources: International Renewable Energy Agency (IRENA). 2014. *Ocean Energy: Technology Readiness, Patents, Deployment Status and Outlook.* Abu Dhabi; IRENA. 2020. *Fostering a Blue Economy: Offshore Renewable Energy.* Abu Dhabi; IRENA. 2020. *Innovation Outlook: Ocean Energy Technologies.* Abu Dhabi; and IRENA. 2021. *Offshore Renewables: An Action Agenda for Deployment.* Abu Dhabi.

Conclusion

Of the main types of MRE platforms reviewed, three are already widely used, two or three appear to be ready for more widespread commercial use, while the others are still in development. Each technology has a different range of benefits and challenges and is more suitable for consideration in various parts of the ocean. The research evidence base is illuminating these differing opportunities all the time. In the next chapter, recent deployments of some of the more promising MRE technologies are reviewed.

CHAPTER 5

Marine Renewable Energy in Action: Innovation and Risk Analysis

Having studied some of the characteristics of MRE platforms—the cornerstone of any Marine Aquaculture, Reefs, Renewable Energy, and Ecotourism for Ecosystem Services (MARES) project—this chapter examines several existing deployments, and also explores some of the environmental and social risks to be assessed when considering each of the main platform types.

Innovations in Marine Renewable Energy

Wave Power

MRE platforms have been developing for some time. The world's first grid-connected wave power station, for example, has been active off the West Coast of Australia since 2015.[102] Produced by Carnegie Wave Energy, the design was centered around the concept of durability (Figure 13). It operates under water, which provides added protection during large storms. This also means it cannot be seen from the shore, which removes the potential for complaints by local communities about ocean views being disrupted. The buoys are tethered to seabed pump units at a depth of between 25 m and 50 m. Pressurized seawater is pushed through a pipeline beneath the ocean floor to an onshore hydroelectric power station where a turbine generates electricity.

Figure 13: Grid-Connected Wave Power Station, Off the Coast of Western Australia

Source: Carnegie Wave Energy (2015) Reproduced at: https://www.sciencealert.com/world-s-first-grid-connected-wave-power-station-switched-on-in-australia.

---

102 M. Gogh. 2015. World's First Grid-Connected Wave Power Station Switched on in Australia. *Science Alert*. 23 February.

Devices that harness energy from waves come in many forms, such as the one manufactured by Eco Wave Power (Figure 14). This platform was first tested in Gibraltar in 2015 and can operate day and night, all year round.[103] The company suggests that the device only requires waves at an average height of 500 centimeters, which makes it viable in most oceans, most of the time. Should waves become stronger than that, though, the design incorporates a storm protection mechanism that "locks" the power-generating buoys to the dock to protect them from potentially damaging waves.

Figure 14: Wave Power Facility, Gibraltar

Source: Eco Wave Power (2019). Reproduced at: https://www.startupselfie.net/2019/04/11/machine-turns-constant-power-of-waves-into-electricity/.

Offshore Wind

As an example of the confidence running through the offshore wind market, construction of the world's largest offshore wind farm began in May 2022.[104] Located at the UK's Dogger Bank, the wind farm will eventually boast installed capacity of 3.6 GW, delivered equally in three phases of construction. Once completed in 2026, the $3.98 billion Dogger Bank Wind Farm will be capable of delivering power to 6 million homes—that is over 20% of all UK households.

---

103   Startup Selfie. 2019. Machine Turns Constant Power of Waves into Electricity. 11 April.
104   M. Lewis. 2022. Construction of the World's Largest Offshore Wind Farm Has Begun. *Elektrek*. 4 May.

Floating Offshore Wind

Equinor's Hywind Scotland, the world's first floating offshore wind farm, started producing electricity in October 2017 and generates enough electricity for more than 20,000 homes.[105] It covers around 4 km² and has installed capacity of 30 MW, drawn from platforms that use a three-line mooring system and operate in water depths between 95 m and 120 m. Further Hywind deployments are already being ramped up, including 140 km off the Norwegian coast.[106] Notably, this deployment pushes further out to sea, to depths of between 260 m and 300 m. As claimed by the manufacturers, it will also be the first floating offshore wind farm to deliver electricity to oil and gas installations, the kind of cross-support function at the heart of the MARES approach.

Figure 15: Floating Offshore Wind Facility

Source: Global Maritime (2021). https://www.globalmaritime.com/.

The pace of technological change continues in the offshore wind sector, as evidenced by the announcement in February 2022 of a new trial of two-bladed floating wind turbines to be deployed in European waters.[107] With a focus on durability, this unique two-bladed design features a special teetering hinge that separates the shaft and rotor, thus protecting the turbine from potentially harmful loads. The company claims that this design thus makes it applicable for deployment in all sea states, including cyclonic regions.

[105]   P. Hockenos. 2020. Will Floating Turbines Usher in a New Wave of Offshore Wind? *Yale Environment 360*. 26 May.
[106]   A. Memija. 2022. First Turbine Installed at World's Largest Floating Offshore Wind Farm. *offshoreWIND.biz*. 7 June.
[107]   A. Sakharkar. 2022. Unique Two-Bladed Floating Wind Turbines Will Be Installed by 2024. *Inceptive Mind*. 24 February.

Ocean Thermal Energy Conversion

This ocean–energy thermal converter (Figure 16), which produces electricity from the difference of temperature between deep cold and warm surface seawater, first received class approval in 2016.

Figure 16: Ocean Thermal Energy Conversion Facility Concept Design

Source: Korea Research Institute of Ships and Ocean (KRISO) (2021). https://www.kriso.re.kr/eng/.

In August 2022, India's National Institute of Ocean Technology announced that it is constructing an OTEC plant with a capacity of 65 kW in Kavaratti Lakshadweep, off its southwest coast.[108] The total OTEC potential around India is estimated at 180,000 MW, and the Geological Survey of India has explored further OTEC potential through six scientific missions around the Andaman and Nicobar Islands and other offshore domains.

Tidal Power

According to the Water Power Technologies Office of the US Department of Energy, tidal energy has the potential to power over 20 million American homes.[109] However, currently, the global tidal and wave power has mostly remained untapped. In New York, the tidal power has been operational for many years. In 2000, Verdant power introduced their Roosevelt Island Tidal Energy Project, and they consider tidal energy as more reliable than solar or wind power since it is easier to predict how much power the turbine will produce over a long period of time without having to rely on the weather for its energy output.

This tidal energy array (Figure 17), produced by Nova Innovation, was launched in Nova Scotia in 2020 and recently received a further $2.3 million (£2 million) of government funding to support mass manufacture for European and North American markets.[110]

---

[108]   G. S. Kaura. 2022. Ocean Thermal Energy Conversion Plant Being Set Up in Lakshadweep. *Tatsat Chronicle*. 3 August.

[109]   A. Gelfland. 2022. Developers Look for 'Sea Change' in Tidal Power Development. *The Energy Mix*. 13 March.

[110]   Marine Scotland Assessment. Case Study: Nova Innovation – Shetland Tidal Array.

Source: Nova Innovation (2016). Reproduced at: https://marine.gov.scot/sma/assessment/case-study-nova-innovation-shetland-tidal-array.

Identifying and Managing Risks

While the pace of technological development is encouraging, broader consideration must be given to the clear environmental and social risks associated with the deployment of such platforms. As MRE technologies are still in the relatively early stages of development, there are understandably few studies on the impacts of these energy conversion methods. Most studies evaluate only the impacts of prototypes with short operating periods, ignoring the need to scale up, replicate, and commercialize. In some cases, this has caused the construction permits to be postponed or denied due to lack of information about innovation possibilities and impacts, and/or social and political support.

An important social factor in gaining public support for implementation of new technologies is public participation during the planning and the decision-making processes. During this process the advantages and potential negative impacts of the new technologies can be reviewed. Lack of public participation and transparency can lead to public opposition. There is a simultaneous need to ensure sufficient awareness and outreach combined with developing the policies, financial incentives, data collection, and other conditions sufficient to an enabling environment.

Building and operating MRE facilities will create numerous new jobs, reduce unemployment, support numerous other local businesses, while reducing expensive imports and developing climate-resilient, future-proofed, socially inclusive livelihoods, thus improving the local economy, health, and decarbonization objectives. In many cases, some new MRE facilities will produce electricity that can be sold at a reduced price, and with far less pollution than fossil fuel power plants.

But there are still some potential risks associated with the main MRE platform types.

Offshore Wind Risks

A range of potential risks relate to offshore wind projects.

Commercial and recreational fishing sectors can have concerns on the possible negative impacts that offshore wind farms may have on their fishing operations. In one case, a wind farm attracted so many recreational fishermen that commercial fishermen were not able to operate there.[111] Other issues could arise by wind farms potentially disrupting established fishing lanes, by the possibility of their gear being entangled with offshore wind cables and equipment, by directly removing productive fishing grounds, or by indirectly affecting fish behavior.[112]

Yet, these changes may not always be detrimental. Like oil rigs, some studies indicate that platforms can serve as artificial reefs, attracting fish.[113]

Because of the potential visual impacts caused by wind farms, a number of projects have received public opposition (footnote 113). Stakeholders' common concerns are that offshore wind development may reduce property values or will discourage tourism in coastal locations. In the 2017 fall Goucher Poll, 11% of the 671 Maryland residents interviewed stated that they would be "less likely" to have a vacation in the coastal town of Ocean City because of the wind turbines.[114] A study of the Block Island Wind Farm, however, revealed that wind turbines have positive economic impacts on tourism.[115]

Offshore wind farms include several factors that have influence on the nature of the produced visual impact:[116]

- the location and size of the wind farm;
- the size, materials, and colors of the wind turbines;
- the arrangement and distancing of wind farms and their associated structures;
- the location, dimension, and form of ancillary onshore (substation, pylons, overhead lines, underground cables) and offshore structures (substation and anemometer masts);
- visibility for navigation including markings and lights;
- the boats used for transport and maintenance;
- the port, slipway, or pier for boats; and
- access requirements to the coast and proposed road or track access.

Wind farms are somewhat noisy. So, if they are located within hearing distance of local coastal residents, noise pollution (for human populations and marine animals) is likely to be a problem. Other visual impacts include shadow flicker (the effect of the sun low on the horizon shining through the rotating blades of a wind turbine, casting a moving shadow); general visual impact; and their effects on tourism.

---

[111]   *Rhody Today*. 2019. URI survey of fishermen finds varied perceptions of impacts of Block Island Wind Farm. 10 January.

[112]   E. Botkin-Kowacki. 2018. Can Offshore Wind and Commercial Fishing Coexist? *The Christian Science Monitor*. 4 September.

[113]   E. Ciara et al. 2020. Social Impacts to Other Communities that Experienced Offshore Wind. In M. Severy et al., eds. *California North Coast Offshore Wind Studies*. Humboldt, CA: Schatz Energy Research Center.

[114]   Goucher College. 2017. *Goucher Poll: Pole Release #1 – Race Immigration, and Hurricanes*. Maryland.

[115]   A. Carr-Harris and C. Lang. 2019. Sustainability and Tourism: The Effect of the United States' First Offshore Wind Farm on the Vacation Rental Market. *In Resource and Energy Economics*. Elsevier B.V. 57. pp. 51–67.

[116]   A. Wratten et al. 2005. *The Seascape and Visual Impact Assessment Guidance for Offshore Wind Farm Developers*. Enviros Consulting and Department of Trade and Industry.

Another impact of offshore wind farms can be on indigenous populations near the areas where farms are constructed. Tribes can have significant and long-standing attachment with the land and oceans used for offshore wind energy development. They can use the regulatory frameworks to defend their heritage.

Wind farms are well known for killing and injuring birds, with one author calling them "bird death traps." Clearly, this is one risk that will be considered in environmental assessments. In areas where endangered species of birds occur, this could result in blocked development even though public support for the project may be high.[117]

Ocean Thermal Energy Conversion and Tidal Currents Energy Impact and Risks

The main impacts of OTEC and tidal technologies are deep-water discharges, which can alter the trophic network and the potential mortality of small organisms entrained in the seawater and passing through the OTEC system. OTEC discharges can change the amount and size of the species that are distributed in the areas near the discharge points.

Further, increases in the noise level created during the construction and establishment of infrastructure may cause the absence of key species, colonization, decrease in population, etc. Installation of ocean–energy arrays can serve as barriers that restrict the movement of the species, resulting in changes in their migratory paths. When extracting energy during tidal currents, the turbines used increase the underwater noise, causing potential tissue damage and stress to fish, marine mammals, and birds.

OTEC plants can be open-cycle (seawater as working fluid), closed-cycle (external working fluid), and hybrid. Further, platforms for OTEC plants can either be offshore or onshore (land-based or near-shore). Because each has different construction and maintenance requirements, there are considerable cost variations between the plants. Compared with land-based plants, the floating platform installation utilizes less land area but requires submarine cables to bring the electricity to land. As a result, the cost of maintenance is higher. An offshore OTEC plant requires marine cables to transmit electric power. This can affect the surrounding environment and increase the capital cost for its construction. The short-term environmental effects of cables include resuspension of sediments and physical disturbances of the habitat brought about by their installation. Potential long-term effects (i.e., during the operational phase) include heat emission, species colonization, and emission of electromagnetic fields. The impacts of offshore plants can be predicted using information on the construction of oil platforms. This may help with minimizing risks and more accurately calculating costs. Nevertheless, because of the visual impact on the landscape, the presence of the plant could cause social disagreement.[118]

On the other hand, this method of energy supply could strongly increase social welfare and supply with energy many isolated communities that are disconnected from the national electricity grid, thus contributing to overall economic development of the country. One of the factors to encourage the placement of the energy plant is the creation of jobs for the people living nearby. During the construction and establishment of an OTEC plant, recruitment of workers from the area can be done; however, during the operation phase, workers with particular knowledge of plant management and maintenance are critical.

---

117   R. Eveleth. 2013. How Many Birds Do Wind Turbines Really Kill? *Smithsonian Magazine*. 16 December.
118   G. R. Camacho, A. F. Delgado, and E. Mendoza. 2020. A Review on Environmental and Social Impacts of Thermal Gradient and Tidal Currents Energy Conversion and Application to the Case of Chiapas, Mexico. *International Journal of Environmental Research and Public Health*. 17 (21). 7791.

Land and Marine Floating Solar Impact and Risks

Compared to conventional energy sources, solar PV systems offer substantial social and environmental benefits, thus promoting sustainable development.

The social and political impacts of solar panels are not yet well understood as most of the extant research deals with the technical and economic aspects of the evaluation. Because solar technology is still young and has a long lifecycle, little is known about its broad impacts.

Some of the positive social and economic impacts of land-based solar PV include reduced land use in comparison to conventional energy resources; use of unused sites (e.g., deserts); reuse of degraded sites; reduced transmission lines and/or grids; multipurpose and integrated use of existing developments or buildings (such as rooftops, façades); energy supply for decentralized, low density off-grid areas, also in developing countries; diversification in energy market; national energy independency from import, jobs creation, and higher development and education level, etc.

The deployment of floatovoltaics or floating solar photovoltaic (FPV) systems is progressing, with various designs starting to appear in a variety of marine environments. As insight from freshwater floatovoltaics is not readily transferable offshore, the lessons from other marine energy infrastructure are used to emphasize how the marine environment may affect PV, how the floatovoltaics have both positive and negative impacts to the environment, and the possible societal response.

One of the major advantages of floating solar panels is that they do not occupy valuable land space, allowing it to be used for other purposes, like farming or construction. A lot of these installations can occupy unneeded space on bodies of water, including wastewater treatment ponds, hydroelectric dams, and drinking water reservoirs.

Solar panels are durable and can operate under high temperatures. But just like other electronics, higher temperatures result in lower power outputs. When temperatures rise, solar panel performance tends to decrease. The cooling effect of the bodies of water that host floating solar arrays improves the performance of solar panels by 5%–10%; this results in considerable cost savings over time.

Floating solar panels can play a significant role in contributing to healthier environments. Water not only cools solar equipment in floatovoltaics, but it also works the other way around. The structure of the floating solar panel provides shade to the body of water, which lessens evaporation from these ponds, reservoirs, and lakes. In freshwater or the ocean, the floating structure may attract fish, which would be a plus for recreational fishermen. However, the shade over a large shallow area would likely cause the lake or seabed communities to change somewhat, with less plants being able to survive.

Floating solar panels generate clean and renewable electricity. Its utilization contributes to the reduction in the emission of greenhouse gases and other pollutants into the atmosphere, leaving a favorable effect on both the environment and human health.

FPV systems also have some disadvantages. Its installations may entail additional costs compared to conventional types of solar panel installations. This is a relatively new technology, thus it usually costs more because it needs specific equipment and more in-depth installation skills.

Most floating solar installations are large scale and supply power to large communities, utility companies, businesses, or municipalities. Floating solar farms are a recent development. They gained traction in 2018, especially in countries with high population size and with competing uses for limited amount of available land.

Also, there are site-specific impacts of the floating solar power plants on the environment and surrounding communities. The construction involves series of activities like site preparation (e.g., grading and leveling) and installation and commissioning of infrastructure (such as the floats, PV panels, inverters, transformers, access road, and transmission lines). These activities could produce air and noise emissions, impact the aquatic habitats, and pose health and safety hazards to workforce and community. However, given the short duration of the construction phase (i.e., a year) and the proposed mitigation measures, these impacts are not expected to be significant.

Some of the key mitigating measures include the use of dust suppression techniques, use of personal protective equipment, traffic management, fencing of the project site, and continued engagement with stakeholders. The PV panels will entail ongoing maintenance during the operation phase, including regular washing of the panels to remove dust and other debris. No chemicals are needed in this process. Personnel will still be required to wear appropriate personal protection gear.

Installation of floating PV modules and anchoring may restrict access to fishing grounds within a reservoir, which is another possible impact. The construction phase could pose several potential health and safety concerns to the community. This covers community health and safety issues that are both influx-related and not related to influx concerns. The construction of project components may expose the local community to health and safety concerns, such as those related to noise, dust, and traffic safety. The impact of noise and dust can be referred to in the following sections on environmental impact assessment. The floating solar structure on the lake may attract the people from within the area or tourists who want to see the project area and observe the system. This may lead to drowning as the people would like to access the lake due to their curiosity to the floating solar modules. The influx of visitors could result in the disturbance to the daily activities of the local community who live near the project area.[119] The number of transportation trips during construction of the project can also be increased and can impact road traffic because of congestion and emissions implications. To minimize the potential for any negative impacts, relevant mitigation measures should be introduced during project construction and operation.

Despite the potential risks, solar energy is a much cleaner source of energy than the burning of fossil fuels as it emits very little pollution into the air. Cities or areas that consider employing solar energy to power the buildings would enjoy a cleaner air quality, which can improve the health of the citizens and workers in the area.

Moreover, island-based solar initiatives can have positive social benefits as a result of decreased heating costs of low-income individuals, reduced reliance on public utilities, job creation, and improved environment from reduced pollution.

Marine Bioenergy Impact and Risks

Bioenergy projects affect the communities in which they operate in several ways. This can range from better water quality to job creation in economically depressed areas. Some use land for feedstock for bioenergy based on dedicated field production (such as energy crops) or residues from agricultural production. Some agricultural

---

[119] Government of Viet Nam. 2018. *Initial Environmental Examination and Social Report: Da Nhim - Ham Thuan - Da Mi Hydro Power Joint Stock Company Floating Solar Energy Project* (prepared for ADB).

areas are marginal for food production, and bioenergy production could improve these marginal lands. However, in some cases, the production of energy crops may have a negative impact on food security.

The World Health Organization reported that one of the top causes of death for children under five years old in poor countries is acute respiratory infections. These infections can be triggered by insufficient ventilation systems to evacuate the buildup of indoor air pollutants from the use of biomass as a cooking fuel. Hence, bioenergy can have a negative impact on air quality. The use of fertilizers and pesticides can also affect water quality. To improve the quality of life of the people who will benefit from bioenergy projects, a proper environmental impact assessment is often necessary prior to its implementation. For instance, the use of briquettes has been investigated as a low-technology, affordable fuel that could be used in developing countries to improve cooking fuel efficiency and enhance indoor air quality.

The production of marine biomass like microalgae for bioenergy and/or biofuel has emerged as a promising renewable energy source. Several research and development activities are still in progress. The National University of Ireland, Galway has successfully developed a demonstration case for offshore applications. According to the findings, the use of marine biomass, if commercialized, could potentially be as significant and comparable to existing land-based forestry and agricultural energy crops.[120] However, literature and knowledge about the social impacts of this new technology are still lacking.

Accelerating Progress

Despite these risks to be managed, with proven technologies escalating up the technology readiness scale rapidly, there is much room for positivity. However, there is still a long way to go. As shown in Figure 18, electricity generation from marine technologies only increased an estimated 13% in 2019.[121] This is significantly above the levels of the previous 3 years, and increased further in 2020, recording an estimated 33% increase, mainly owing to Denmark's capacity increase of 200 MW. However, this is still far below the annual growth of 23% required to meet the International Energy Agency's sustainable development targets by 2030.

The most up-to-date report from the International Renewable Energy Agency notes that to achieve its goal, it would require an average of 1 GW of capacity additions annually until 2030. The report also comments that many demonstration and small commercial projects remain expensive because "the economies of scale necessary for significant cost reductions have not yet been realized," noting that additional policy support for research, development, and demonstration is urgently required to enable the cost reductions that come with the commissioning of larger commercial plants.

120   L. M. L. Laurens. 2017. *State of Technology Review – Algae Bioenergy*. IEA Bioenergy Technology Collaboration Programme.
121   IEA. *Ocean Power Generation in the Net Zero Scenario, 2000-2030*.

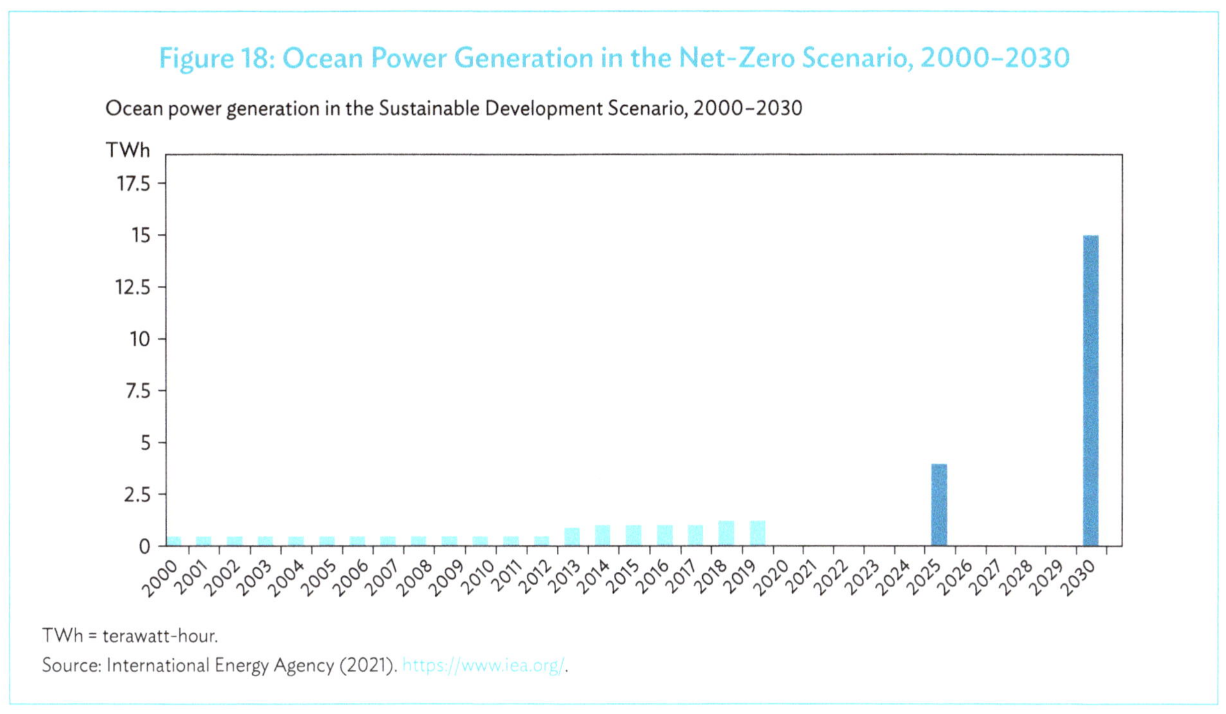

Figure 18: Ocean Power Generation in the Net-Zero Scenario, 2000–2030

TWh = terawatt-hour.

Source: International Energy Agency (2021). https://www.iea.org/.

There are many reasons associated with this perceived lack of progress within the marine and maritime sector, two of which are the following:

- First, not enough may be being done to break out of siloes and look at co-location, integration, co-development, and support of opportunity.
- Second, and particularly when considering small island and/or large ocean states, one of the biggest barriers to investment is that there is only limited need for renewable energy near the generation sites, so there is not a compelling enough commercial driver for investment.

These issues will be addressed in Chapter 7, where the kind of multifunction initiatives that the MARES project has been designed to find will be highlighted. In the next chapter, an equally compelling investment story, linked to the development of marine hydrogen, will be explored. In this scenario, new markets are created where excess MRE is converted to exportable hydrogen. This may unlock the financing required to accelerate MRE development and secure a key cornerstone of nations' Blue Economies.

Conclusion

Offshore wind is well-established in many locations. Floating solar is steadily growing. Other MRE deployment has been progressing in a slower manner than many had hoped for, though many interesting methods are emerging across the technology landscape. There are a number of known benefits and risks associated with each MRE type. While each of them is quite new, knowledge of their environmental impacts is increasing all the time and must be extended. In the next chapter, the development of marine green hydrogen as the possible catalyst for the sector is explored.

CHAPTER 6

Marine Hydrogen: Harnessing Market

This chapter will illustrate how the emerging marine hydrogen opportunity potentially creates a whole new horizon for the MRE sector, by allowing large ocean states to use their marine space to convert energy to electricity, and then to export that to an entirely new market. It will also explore how both policy and commercial drivers are demanding alternative marine fuels, which creates new possibilities in this realm.

Introduction

Having reviewed the various technological developments emerging that make further scaling of MRE possible, the succeeding sections will discuss the second distinct key opportunity that makes the MARES timely.

With the global energy mix, contribution of ocean energy is predicted to remain much smaller than it could be in the next few years. Therefore, further ways to catalyze the sector, and new compelling narratives, are needed to secure investment to support the scaling of the available technologies. One opportunity for such progress is created by the emergence of marine green hydrogen.

This new, portable energy storage solution could allow large ocean states to utilize their significant coastal and marine areas not only to meet local energy demands, but also to convert energy to electricity. Nations can then export the electricity to an entirely new market (known as a "power-to-x" capability). Encouragingly, this technical ability also aligns with new market possibilities. Both policy and commercial drivers are demanding alternative marine fuels, which creates significant new commercial possibilities for marine green hydrogen. This is the second potential opportunity—financial investment will flow into regenerative marine activities because there is a promising export market model that simply did not exist before.

Explaining Green Hydrogen

Electricity is generally produced from MRE options presented in the previous chapter. To replace fossil fuels used in several sectors (including transport, cooling, and industry), electricity energy can now be converted into hydrogen, which can thus be viewed as another form of energy carrier.

Hydrogen can be produced using various methods: from use of conventional energy sources like natural gas or coal; from nuclear energy; and, most importantly, from renewable energy sources like wind, solar, and biomass, as shown in Figure 19.

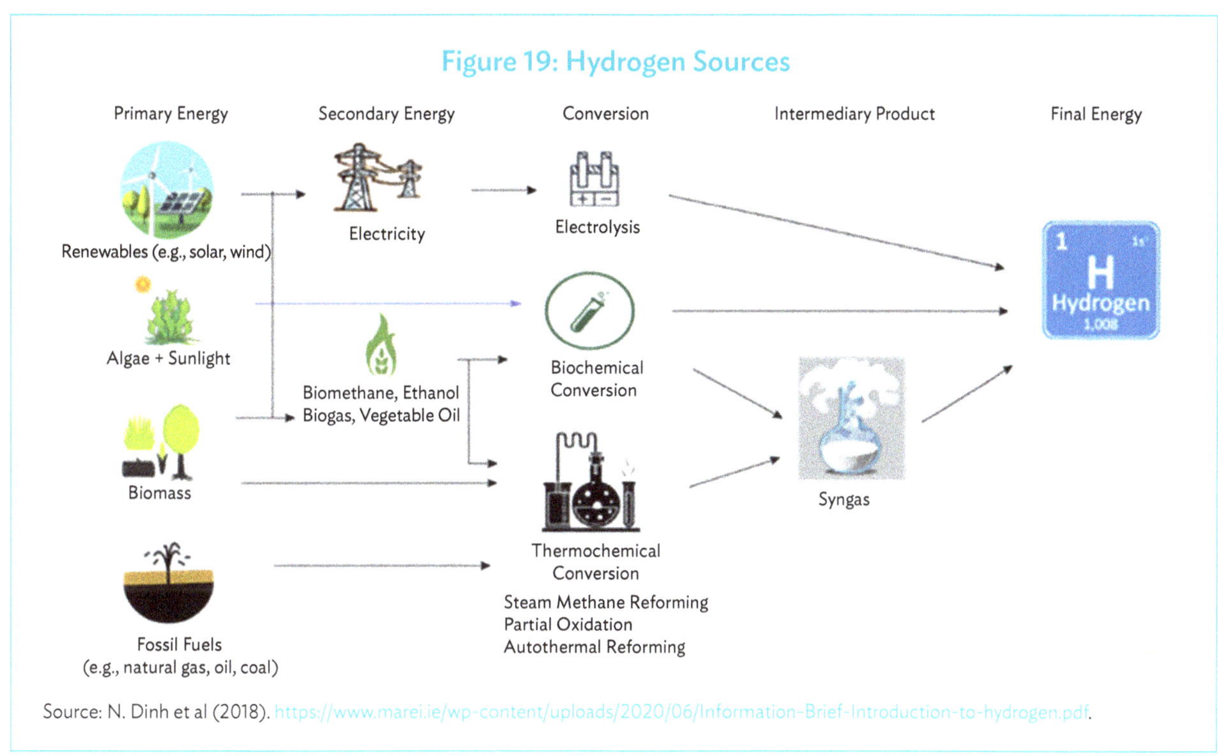

Source: N. Dinh et al (2018). https://www.marei.ie/wp-content/uploads/2020/06/Information-Brief-Introduction-to-hydrogen.pdf.

These fuel sources can be used to produce hydrogen through electrical, thermal, and photochemical processes.[122] The most common and commercially available method for production of hydrogen is steam methane reforming and electrolysis of water.[123] Currently, 98% of hydrogen production is from fossil fuels, the majority using the steam methane reforming method of natural gas.[124] However, green hydrogen production from renewables has been gaining increasing prominence in recent years because of the promising role of hydrogen in potentially addressing the (i) intermittency of renewable energy resources, (ii) high dependence of electricity generation on imported natural gas, (iii) electricity grid issues, and (iv) decarbonization goals globally.

At present, water electrolysis is the most basic industrial process to produce the purest form of hydrogen and its significance is expected to grow in the future. Water electrolysis is the process of decomposing water into hydrogen and oxygen using electricity. The process of conversion of water into hydrogen requires both thermal as well as electrical energy.[125]

[122] S. Sharma and S. K. Ghoshal. 2015. Hydrogen the Future Transportation Fuel: From Production to Applications. Renewable and Sustainable Energy Reviews. 43 (12). pp. 1151–1158; and S. Z. Baykara. 2018. Hydrogen: A Brief Overview of Its Sources, Production and Environmental Impact. *International Journal of Hydrogen Energy*. 43 (23). pp. 10605–10614.

[123] V. Pushpoth, V. N. Dinh, and E. McKeogh. 2018. *Local Energy Storage for Offshore Windfarms*. School of Engineering, University College Cork. 20 April.

[124] I. Dincer and C. Acar. 2015. Review and Evaluation of Hydrogen Production Methods for Better Sustainability. *International Journal of Hydrogen Energy*. 40 (34). pp. 11094–11111.

[125] S. O. Amrouche et al. 2016. Overview of Energy Storage in Renewable Energy Systems. *International Journal of Hydrogen Energy*. 41 (45). pp. 20914–20927.

Hydrogen produced with electrolysis using the electricity made from renewable energy sources has low global warming potential as well.[126] The production of hydrogen from wind-based electrolyzers is the most effective way of ensuring minimum pollution. The most important advantage of electrolysis of water is the production of extremely pure hydrogen, with the only by-product being oxygen. The major drawbacks at the moment are the investment cost and low overall efficiency.[127]

However, there is potential to create a new global commercial opportunity where excess MRE is converted to exportable hydrogen, which could unlock the financing required to speed up adoption of these technologies and secure a key regenerative cornerstone of many nations' Blue Economies.

Green Hydrogen in Action

The world is already witnessing ways in which green hydrogen is being successfully introduced into the energy mix in innovative ways. Five examples are presented here: (i) the hydrogen production plan for Valentia Island, Ireland (Box 2); (ii) the green hydrogen production barge concept design (Box 3); (iii) the Orion Clean Energy Project (Box 4); (iv) the hydrogen power pilot project in Orkney Islands, Scotland (Box 5); and (v) the Building Innovative Green Hydrogen systems in an Isolated Territory (BIG HIT) project (Box 6).

Box 2: Hydrogen Production Plan for Ireland's Valentia Island

The Valentia Island, which is located off the coast of County Kerry, Ireland, is an example of a small community that seeks to produce hydrogen for their own use. The Valentia Island Sustainable Energy Community aims to establish an energy cooperative and has produced an ambitious energy master plan. One of the key strategic goals that is detailed in the plan is the production of hydrogen to achieve energy sustainability.

The development of GenComm's Decision Support Tool will be very appropriate for this project and will be very helpful in in making hydrogen production a reality for such a remote and small community, which has a history of innovation and technological firsts.

Source: NLA International.

---

126   U. S. Department of Energy, Office of Energy Efficiency & Renewable Energy, Hydrogen and Fuel Cell Technologies Office. Hydrogen Storage.
127   M. Reuß et al. 2017. Seasonal Storage and Alternative Carriers: A Flexible Hydrogen Supply Chain Model. *Applied Energy*. 200. pp. 290–302.

Box 3: Green Hydrogen Production Barge Concept Design

The figure introduces a solution for local hydrogen production within a typical medium-sized commercial port or harbor in a manner that is readily installable, with minimal infrastructure, to support the adoption of hydrogen as a marine fuel. The concept has been developed by the ABL Group who has validated the modeling assumptions, such that other United Kingdom ports can be evaluated for suitability and roll out of the concept. The project objectives are also to develop barge concept designs to suit a range of demand sizes; identify how constraints, costs, and capacity vary with the size of the barge and required output; and further develop a selected concept option detail to prepare for a future demonstration phase.

ABL Group, including group company Longitude Engineering, in a consortium with Green Hydrogen Solutions and Poole Harbour Commissioners, are working on the concept development and feasibility study of a moored barge that would generate, store, and supply hydrogen to vessels bunkering at the same port. The port will store hydrogen that is produced using onshore renewables via electrolysis. Poole Harbour will be used as a case study to show the concept's environmental and commercial viability.

Green Hydrogen Barge Concept Designed by the ABL Group

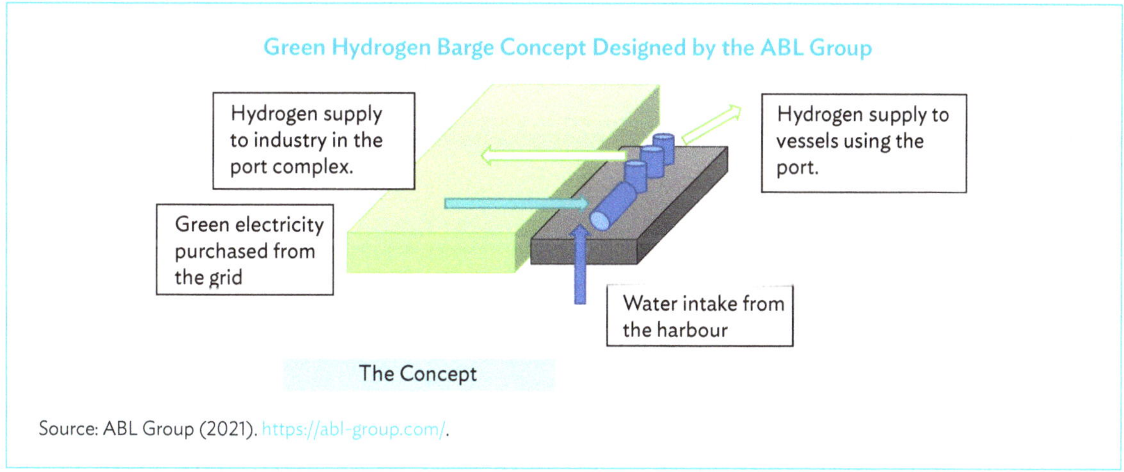

Source: ABL Group (2021). https://abl-group.com/.

Source: V. N. Dinh and E. McKeogh. 2018. Offshore Wind Energy: Technology Opportunities and Challenges. In *Lecture Notes in Civil Engineering*. Vol. 18. pp. 3–22. Switzerland: Springer.

Box 4: ORION Clean Energy Project—Shetland Electrification and Hydrogen Hub

ORION is a strategic framework that aims to shape Shetland as a world leading green energy island. Its vision is transformative and aims to place Shetland at the heart of cutting-edge clean energy developments, create a highly skilled workforce in the isles, and provide economic security for years to come.[a] Shetland has played a prominent role in the United Kingdom's energy sector since the 1970s, with significant infrastructure built at the Sullom Voe Oil Terminal, the Shetland Gas Plant in the North Mainland, and at Lerwick Harbour in the central part of the isles. But as the world shifts away from fossil fuels in the face of climate change, Shetland must adapt too. That is where ORION comes in. As a catalyst for change, ORION will encourage collaboration with Shetland stakeholders and projects, with a net-zero focus to help generate clean, affordable power; eradicate fuel poverty; protect the environment; and provide job opportunities throughout the isles' supply chain.[b]

The production of hydrogen in Shetland could open a new economic market using local hydrogen, as well as exporting it at national and international markets through the existing connections to the United Kingdom mainland and Europe. This new industry would put to use the long-standing local energy knowledge and expertise of the current workforce and create new jobs in support of the transition toward clean energy technologies, away from oil and gas.

[a] Orion Clean Energy Project.
[b] P. D. O'Kelly-Lynch et al. 2020. Offshore Conversion of Wind Power to Gaseous Fuels: Feasibility Study in a Depleted Gas Field. *The Journal of Power and Energy, Part A of the Proceedings of the Institution of Mechanical Engineers*. 234 (2). pp. 226–236.
Source: Compiled by authors.

Box 5: Piloting Big Technologies on a Tiny Scottish Island

The Orkney Islands, which are located just above mainland Scotland, may appear to be at the end of the earth, but the hydrogen power projects started there could help usher in a new era for marine transportation. A consortium consisting of government, business, and community organizations is developing programs to test hydrogen for its potential as a fuel for a ferry that passes between the Orkney mainland and the tiny island of Shapinsay. This initiative could be the start of a huge global shift from diesel power to renewable energy for short- and medium-haul ships.

Orkney is an archipelago consisting of 70 islands, with a population of only 22,000; yet, it has abundant wind, wave, and tidal resources—all of which are being used to generate energy. Alternative sources of energy on the island generated more than 120% of the electricity consumed locally. Several municipalities there have established community-owned businesses that supply energy locally and to the United Kingdom national grid.

Shapinsay is one such location. It is a small community of only 315 people; yet, it stands out as a model of how local authorities can generate their own energy on behalf of their residents. If many other communities around the world followed suit, it could have significant impact. David Hibbert, a technical superintendent of the Orkney Harbour Authority that operates nine ferries between the islands, says, "It is a dream that communities can generate their own energy to run their own transport. You do not have to be dependent on fossil fuels and the multinational energy suppliers."

Source: S. Hamm. 2021. Piloting Big Technologies On a Tiny Scottish Island. *Technomy*. 9 November.

Box 6: From Hydrogen Islands to Hydrogen Valleys

Orkney Islands initiated the Building Innovative Green Hydrogen systems in an Isolated Territory (BIG HIT) project that utilizes excess energy from the wind turbines to produce hydrogen. Before long, they were not only able to use the windmills during peak demand, but they were also able to use them beyond electricity, powering vehicles, and heating homes.

"Like many islands, Orkney has had challenges with energy," says Nigel Holmes, manager of the BIG HIT project with the Scottish Hydrogen and Fuel Cell Association. "They have had to import energy as liquid fuels—liquid propane gas, petrol, diesel, and heating oil. That means the cost of fuels on Orkney is higher than on the mainland, and cash is flowing out of the local economy." Now, hydrogen is heating community buildings and powering vans on the island.

Source: D. Keating. 2020. Hydrogen Islands to Hydrogen Valleys. *Euractiv.* 17 June.

Fueling the Marine Transport Sector

One of the economic drivers behind the potential for hydrogen and associated storage options lies in the maritime sector's urgent need to decarbonize. Increasingly stringent emissions reduction targets are being introduced into policy and regulatory schemes around the world, and to meet these requirements the shipping, fishing, cruise, and associated sectors are seeking alternative fuel sources to replace being powered by fossil fuels. Hydrogen could be used to form synthetic fuels such as ammonia (combined with nitrogen) and methanol (combined with carbon monoxide), which are free of harmful emissions. These fuels could not only help decarbonize the maritime sector, but also—for those nations that are able to produce them—decrease the reliance on imported fuels, thus also strengthening local economies. Some initial examples of hydrogen-enabled maritime operations are featured below.

Sea-Going Hydrogen-Powered Ferry

Europe's first sea-going ferry powered by hydrogen fuel cells is being developed by the Scottish-led HYSEAS III project. It is a double-ended sea-going passenger and car ferry, utilizing the hydrogen-powered drive train and thereby running completely emission-free, and will have capacity for 120 passengers and 16 cars or two trucks. The designs are being provided by ABL Group, with the first renderings shown in Figure 20.

ABL Group's company Longitude was contracted by Caledonian Maritime Assets Limited to undertake the design for this ferry. The vessel is of a Ro-Ro design to maximize overall efficiency. By utilizing fuel-cell technology, the vessel is "zero-emission" and will utilize the local supply of green hydrogen. Consideration is also given within the design to minimize emissions and waste during the build and disposal of the vessel. Propulsion options include the electric Voith Schneider Propeller (eVSP) unit with integrated electric motor, again enhancing efficiency.

Longitude, as the first of its type in the UK, is collaborating closely with Class and the UK flag authority to develop the design philosophy for the novel aspects and ensure the highest standards of reliability and safety for the vessel. Further information regarding hydrogen production and applications in the UK could be sourced from the green hydrogen policy report recently published by Renewable UK.[128]

---

[128]   C. J. Jepma. 2015. *Smart Sustainable Combinations in the North Sea Area.* Groningen: Energy Delta Institute.

Figure 20: First Renderings of the Hydrogen-Powered Vessel Designed by the ABL Group

Source: Caledonian Maritime Assets Ltd (2021). https://www.cmassets.co.uk.

Hydrogen-Power Tug

A concept for a 65-ton, electric, harbor tug (Figure 21) has been developed by the ship design and marine consultancy firm, OSD-IMT (an ABL Group company), and polymer electrolyte membrane, Fuel Cell Market Leader Nedstack.[129] The electric power to run the propulsion motors is generated by hydrogen fuel cells onboard, making the tug completely emission-free. The hydrogen is stored in quickly and safely exchangeable 20-feet ISO containers, with 2-day to 4-day endurance in normal harbor operation. This design is meant to be scalable in terms of performance to larger and smaller tugs (footnote 129).

Figure 21: Concept for a 65-Ton Harbor Tug Powered by Hydrogen

Source: Nedstack (2019). https://nedstack.com/.

---

[129]   C. J. Jepma and M. van Schot. 2017. *On the Economics of Offshore Energy Conversion: Smart Combinations—Converting Offshore Wind Energy into Green Hydrogen on Existing Oil and Gas Platforms in the North Sea.* Groningen: Energy Delta Institute.

Conclusion

The marine green hydrogen opportunity potentially creates a whole new blue horizon by allowing large ocean states to utilize their coastal and marine areas not only to meet local energy demands, but also to convert energy to electricity, and then to export that to an entirely new market (a "power-to-x" capability).

Both policy and commercial drivers are demanding alternative marine fuels—e.g., which creates significant new commercial possibilities in this realm—that will attract new financial investment.

If they can be realized, these new strategic drivers potentially create a very different future for coastal and island states with large EEZs. In the past, commercial investment in MRE options has been challenging for island states, as local demand could not promise healthy returns on the investment required to implement renewable options. As marine green hydrogen creates new opportunities for export potential, there are many new possibilities.

With this in mind, the real potential of renewable-powered multifunction projects in island states can be explored. That theme is discussed further in the next chapter.

CHAPTER 7

The MARES Multifunction Approach

While the introduction of marine renewable energy (MRE) options is, on its own, of great value to coastal and island states, the Marine Aquaculture, Reefs, Renewable Energy, and Ecotourism for Ecosystem Services (MARES) approach pushes for more, especially exploring how regenerative ocean power methods can be aligned to other marine and maritime functions. This chapter shows how regenerative, multifunction marine projects, which have MRE at their heart, could show promise greater than the sum of their parts.

New investments beginning to pour into the blue economy and the potential to produce exportable products drawn from MRE present two significant opportunities for coastal and island states. Against that positive backdrop, there is a third set of opportunities that the MARES project is attempting to catalyze. This is to implement a much broader set of multifunctional regenerative marine projects that make the most effective use of marine areas, with the greatest positive benefits for local communities and the minimum environmental impacts. That is the MARES approach, but what are the added benefits when such components all come together at the same time? How do sector impacts become even more regenerative when they are powered by MRE platforms? And what enabling factors need to be in place? Some case studies are beginning to answer some of these questions.

The Role of Microgrids

First, there is a need to consider one of the key elements required for MARES projects to work on islands and to support island communities. A typical problem affecting "island communities" is that they often are disconnected from their national power grids and must therefore seek other options to provide the power needed to sustain their operations.

The Philippines, for example, has an urgent need to address the unserved and underserved communities residents on some of its 7,000+ islands. To meet this challenge, the government set out a strategy of 100% electrification by 2022. Microgrids were central to that objective. Republic Act No. 11646, also known as the "Microgrid Systems Act," allows microgrid systems providers to operate in any area where there is no electricity access at all or where the power connection does not provide 24/7 electricity supply. The Philippine National Electrification Administration reports that there are 12,672 underserved on-grid barangays, while there are 5,262 identified off-grid and remote areas for electrification as of November 2020.

Looking more broadly at Southeast Asia, there are up to 45 million people (of the 385 million total population) in the region lacking access to electricity, and many islands with insufficient and expensive fossil-based power generation.[130] From this data, a spatial analysis has been completed to identify the location of island communities

---

[130] P. Bertheau et al. 2020. Assessment of Microgrid Potential in Southeast Asia Based on the Application of Geospatial and Microgrid Simulation and Planning Tools. In O. Gandhi and D. Srinivasan, eds. *Sustainable Energy Solutions for Remote Areas in the Tropics.* Switzerland: Springer. pp. 149–178.

with the potential for microgrid distribution of renewable energy. The analysis concludes that the potential is substantial (Figure 22). Up to 1,900 islands and shoreline areas, with a total population of 7.9 million, represent a potential target for microgrid implementation. Given the remote nature of these communities, renewable energy is likely to be the generation source of first choice.

Figure 22: Island Locations in Southeast Asia with Renewable Energy Microgrid Potential

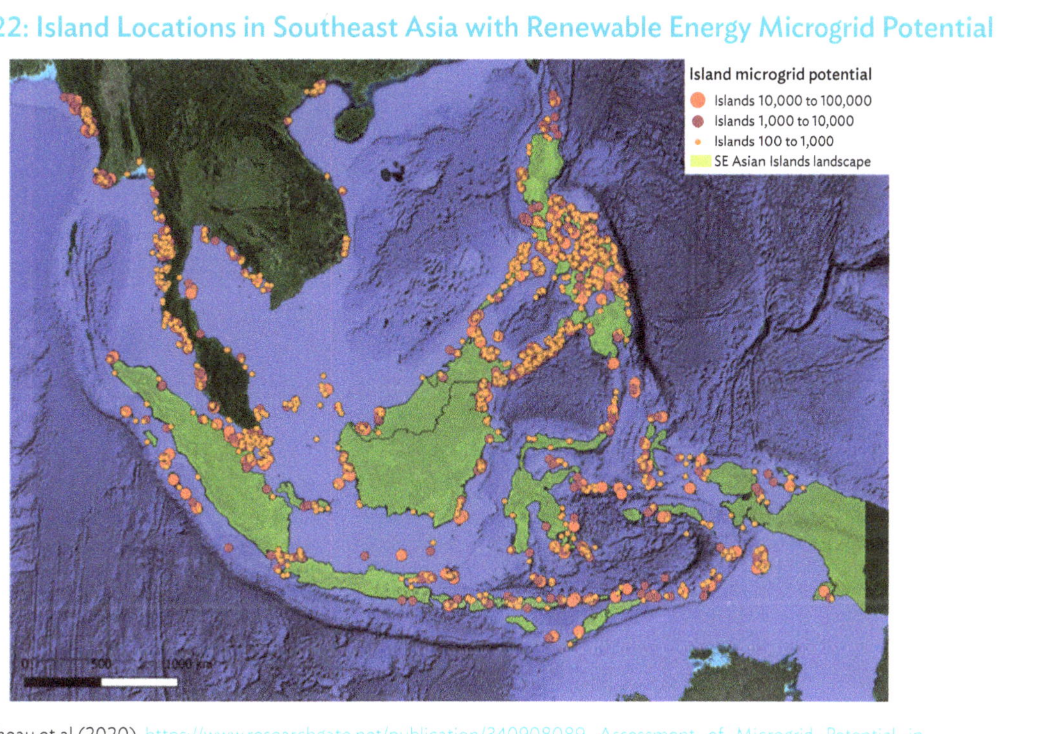

Source: P. Bertheau et al (2020). https://www.researchgate.net/publication/340908089_Assessment_of_Microgrid_Potential_in_Southeast_Asia_Based_on_the_Application_of_Geospatial_and_Microgrid_Simulation_and_Planning_Tools.

Given this conclusion, modeling of island and isolated communities has been of particular interest to the academic community. Three different case studies for tropical islands have been modeled by Bertheau et al. (footnote 130). The models represent the case of small islands (population < 1,000), medium islands (1,001 < population < 10,000), and large islands (population > 10,000). They demonstrate that it is important to first understand the demographics, which translate to an understanding of the likely demand profile.

Next, an analysis of renewable energy resource potential is determined. In many cases, solar PV is found to be highly attractive because of the irradiance profiles of the areas under study. The preliminary data is then examined using simulation tools to determine various metrics, including the expected levelized cost of energy (LCOE).

A variety of tools have been suggested for this analysis, including the Simplified Planning Tool for Hybrid Systems (from the Reiner Lemoine Institute, Excel-based); Offgridders (Open source, Python 3-based); and commercial product Hybrid Optimization of Multiple Energy Resources. The application of these tools on three islands ("Small" population < 1,000; "Medium" 1,001 < population < 10,000; and "Large" population > 10,000), concludes that the microgrid and/or hybrid generation, including renewables, not only improves energy security, but also results in an overall lower LCOE (Table 4). For example, looking at the mid-sized island group, with a typical population of 3,000–4,000, the simulation has shown that savings can be realized over strictly fossil fuel-based systems.

Table 4: Levelized Cost of Energy Case Study Results for Small, Medium, and Large Islands

Size of System	Type of System	Energy Source	Diesel Capacity kW	RE Share	Solar PV kWp	Battery kWh	LCOE k$/Whr	# of Diesel	Fuel k liters/ year	Fuel Cost k$/yr	O&M k$/yr
Small	Baseline	Diesel	25	0%	0	0	0.59	1	33.8	27.1	2.2
p<1k	Hybrid	D+RE	25	31%	20	28	0.35	1	14.2	11.3	0.7
Medium	Baseline	Diesel	276	0%	0	0	0.36	2	326	260	24.8
1k<p<10k	Hybrid	D+RE	276	32%	266	180	0.28	2	192	154	9.4
Large	Baseline	Diesel	3,600	0%	0	0	0.34	3	4,191	3,353	350.4
10k<p	Hybrid	D+RE	3,600	48%	7,200	1,500	0.26	3	2,071	1,657	121.4

D = diesel, k = thousand, kW = kilowatt, kWh = kilowatt-hour, kWp = kilowatt-peak, LCOE = levelized cost of energy, O&M = operation and maintenance, p = population, PV = photovoltaic, RE = renewable energy, Whr = watt-hour, yr = year.

Source: P. Bertheau et al. 2020. Assessment of Microgrid Potential in Southeast Asia Based on the Application of Geospatial and Microgrid Simulation and Planning Tools. In O. Gandhi and D. Srinivasan, eds. *Sustainable Energy Solutions for Remote Areas in the Tropics*. Switzerland: Springer.

For a substantial increase of renewable penetration, and thus potential decrease of power generation costs, it is necessary to implement solar PV with battery storage, given the higher electricity demands in the evening.[131]

The case for microgrids is thus becoming more compelling and more feasible in the case of island states. This will help to prepare the ground for more ambitious MARES multifunction initiatives.

Multifunction Marine Projects—Some Examples

Chapters 8 and 9 will suggest the type of multifunction projects that may be applicable in the selected island states of the Marshall Islands and Palau. Some projects of that nature that are already being implemented in various parts of the world will be highlighted to further underscore potential and feasibility.

Some multifunction approaches are already operating strictly within the MRE sector. In Gran Canaria, for example, the first full-scale hybrid floating wind and wave platform is being developed. The platform, shown in Figure 23, will be able to generate over 5 MW of power from the wind turbine and wave energy converters, and the system allows for grid connection through a new subsea cable.[132]

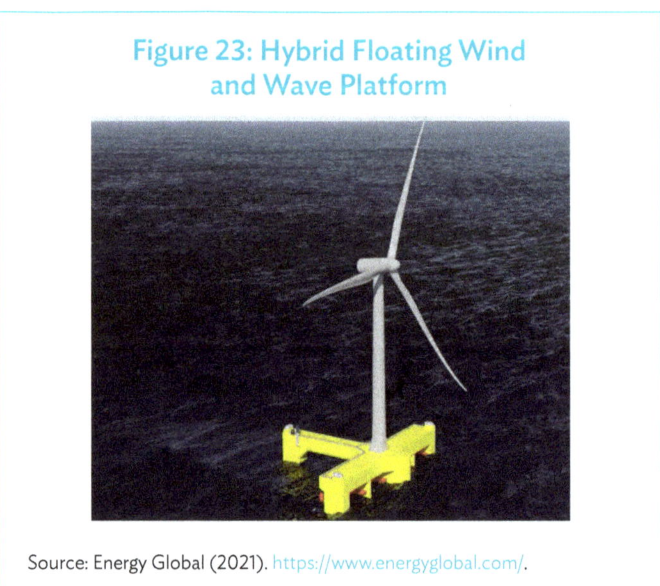

Figure 23: Hybrid Floating Wind and Wave Platform

Source: Energy Global (2021). https://www.energyglobal.com/.

Taking that concept one step further, the floating structure from SINN Power in Figure 24, launched in the summer of 2020, attempts to harness three renewable sources at once—waves, wind, and solar.[133]

131   US Department of Energy, Federal Energy Management Program. 2016. Ocean Energy. *Whole Building Design Guide.* 18 November.
132   *Energy Global.* 2021. Floating Power Plant and PLOCAN Sign Wind and Wave Contract. 29 November.
133   C. Casella. 2020. This Clever Ocean Power Station Harvests Wind, Wave and Solar Energy on One Platform. *Science Alert.* 30 May.

Figure 24: Hybrid Wave, Wind, and Solar Device

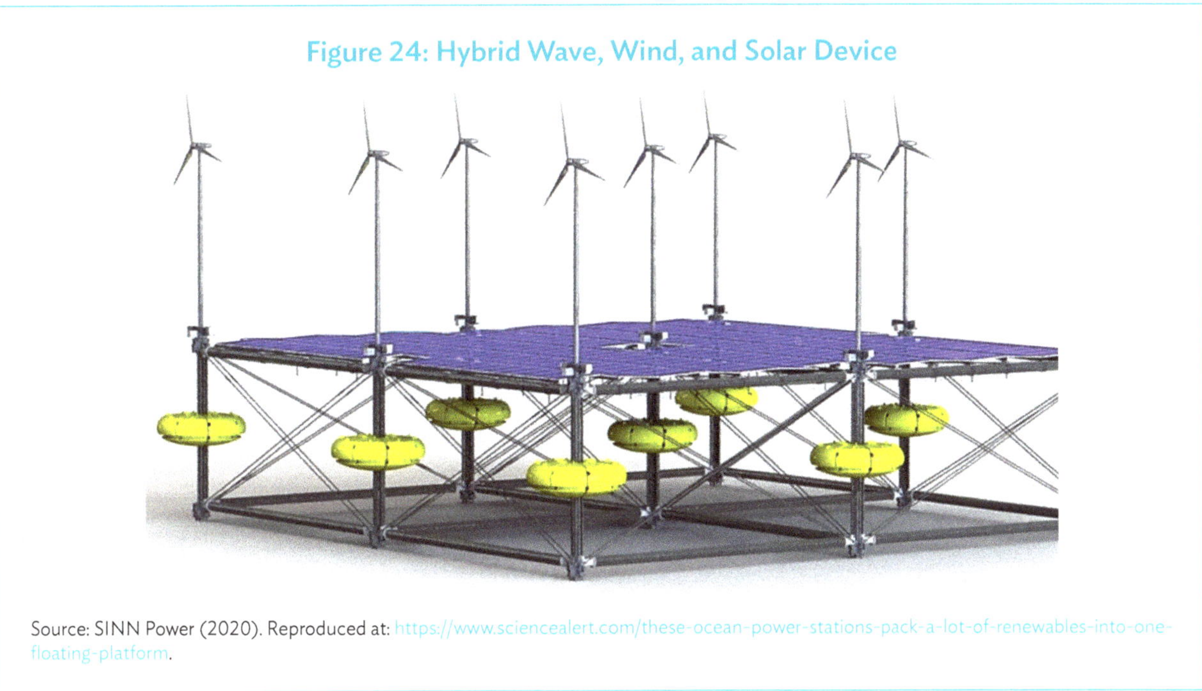

Source: SINN Power (2020). Reproduced at: https://www.sciencealert.com/these-ocean-power-stations-pack-a-lot-of-renewables-into-one-floating-platform.

Renewable energy options are also finding integration paths within the shipping sector. Marine solar panels have been developed that utilize marine-grade steel mount frames to withstand the harshest conditions at sea (Figure 25). They enable shipping companies to add a renewable option to their energy mix, thus improving their cost-effectiveness and environmental footprint.[134]

Figure 25: Ship-Based Solar Panels

Source: Eco Marine Power (2020). https://www.ecomarinepower.com.

---

[134]   Eco Marine Power. Marine Solar Panels and Frames.

Aquaculture projects are already beginning to look at integrating renewable energy solutions (Figure 26). A test project in Scotland looked to develop a clean energy storage system, recharged from onshore renewable energy via a lightweight subsea cable. Utilizing renewable energy in this way will allow fish farm owners to reduce operating costs and cut carbon emissions in the sector.[135]

Figure 26: Marine Renewable Energy Aquaculture Power Project

Source: Ocean Kinetics (2021). Reproduced at: https://thefishsite.com/articles/project-aims-to-boost-the-use-of-renewables-in-aquaculture.

Options to link renewable energy sources and cultivated reefs are also being developed. International NGO, The Nature Conservancy, is working with industry to optimize both marine habitat and clean energy during offshore turbine construction off the coast of New York.[136] Careful material selection and sensitive design choices are being employed to help better support marine ecosystems, effectively maximizing the crossover potential of turbines as artificial reefs.

A more long-standing alignment between energy and coral reefs is the rigs-to-reefs initiative. In 2022, the Government of Brunei Darussalam allocated B\$3 million (\$2.13 million) to increase yields in its capture fisheries, acknowledging the importance of marine ecosystem health by investing in the construction and deployment of artificial reefs. These plans had the ultimate objective of helping to increase the productivity of marine resources by over 50%, with the artificial reef areas providing protected spaces to breed small fish.[137]

---

135   *The Fish Site*. 2021. Project Aims to Boost the Use of Renewables in Aquaculture. 4 October.
136   L. Galst. 2021. Turbine Reefs: Designing Offshore Wind Power to Improve Habitat for Marine Life. *The Nature Conservancy*. 29 November.
137   J. Kon. 2022. Turning the Tide on Reef Damage. *Borneo Bulletin*. 19 June.

This is complemented by the government's continued support of its rigs-to-reefs program. For this approach, redundant offshore oil rig platforms are converted into artificial reefs. This program has been in existence for many years,[138] and it is notable that some rig-to-reef sites have additional value as tourist dive sites.[139]

Perhaps the greatest and most obvious alignment between renewable energy and the tourism sector is to ensure that tourism facilities (e.g., hotels, cruise ships, etc.) are as energy-efficient as possible from renewable sources. There is growing consumer demand for eco credentials, so a strong economic driver is in place. The Elysian retreat in Australia, for example, is one of a number of hotels that are entirely carbon neutral, with each of its seven self-catering units completely powered by hydro and solar power. The company estimates that this enables them to save 39,420 liters of diesel from being burned per year, plus the additional fuel it would have used to transport that fuel to the island via barge.[140]

Pressing humanitarian and community needs can also be addressed through multifunction approaches. For example, researchers from Shanghai Jiao Tong University and Massachusetts Institute of Technology have built a low-cost desalination system (Figure 27) that is powered by renewable energy using sunlight for heat and natural convection for water evaporation.[141] While still at an early stage, removing power costs from the desalination process is an extremely promising step forward in the search for sustainable water security.

Figure 27: Simple Water Desalination System

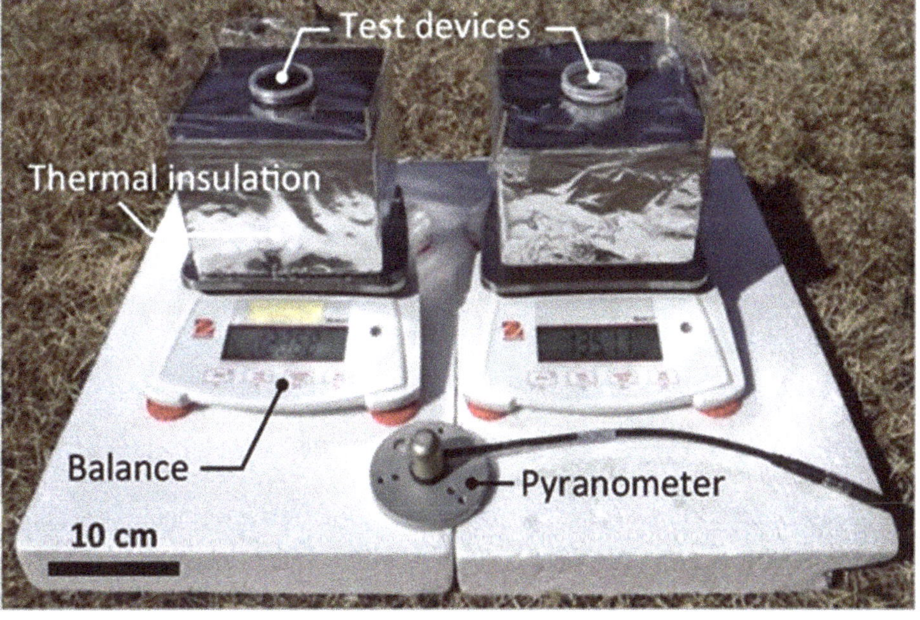

Source: Source: Zhang, L., Li, X., Zhong, Y. et al. Highly efficient and salt rejecting solar evaporation via a wick-free confined water layer. *Nat Commun* **13**, 849 (2022).

138   D. Twomey. 2012. *Artificial Reefs*. Presentation for the CCOP/EPPM Workshop on End of Concession and Decommissioning. Bangkok. 12-14 June.
139   Professional Association of Diving Instructors. Oil Rig Reef.
140   Elysian Long Island Great Barrier Reef.
141   V. B. Ramirez. 2022. This $4 Desalination System Can Meet a Family's Daily Water Needs. *Singularity Hub*. 16 February.

Conclusion

The MARES approach will help us develop a much broader understanding of the sea as a multi-use space starting with three areas: eco- and nature-based tourism, aquaculture, and cultivated reefs. Some examples are already emerging of multifunctional approaches across those sectors, especially aligned to MRE options, and the development of microgrids. But more needs to be done.

The next two chapters will examine the needs for such approaches within the specific states of the Marshall Islands and Palau.

3

CHAPTER 8

In Focus: The Marshall Islands

Having explored the specifics of marine renewable energy (MRE) for island and coastal states and studied the apparent development opportunities in a number of related sectors, the next two chapters pull together some of these strands to see how they might be employed in specific island settings. Some of the more promising MRE options for the Marshall Islands (in this chapter) and Palau (in Chapter 9) will be explored, and how such platforms may provide opportunities for specific multifunction opportunities in those island states will be discussed.

Marine Renewable Energy in the Marshall Islands

The Republic of the Marshall Islands spans over 5 million km^2 and comprises 1,225 islands and islets, including 29 atolls and five solitary low-lying islands.[142] Only 0.01% of the country is land (about 181.3 km^2). The atolls and islands form two parallel chains, the eastern chain (Ratak or Sunrise) and the western chain (Ralik or Sunset). These two chains are part of the Ratak and Ralik Ridges and comprise numerous seamounts, some of which are guyots (flat-topped seamounts).[143]

The Marine Aquaculture, Reefs, Renewable Energy, and Ecotourism for Ecosystem Services (MARES) project team reviewed the national preexisting policies and strategies of the Marshall Islands to prepare the ground for ensuring that project proposals align with these. The Division of Energy within the Ministry of Resources and Development is responsible for National Energy Policy Coordination and some implementation. This includes responsibility for energy efficiency and renewable energy. The Framework for the National Energy Policy includes a 20% increase in energy efficiency and 100% complete renewable energy or zero emissions.

The 2009 Marshall Islands National Energy Policy calls for, "an improved quality of life for its people through clean, reliable, affordable, accessible, environmentally appropriate, and sustainable energy services."[144]

In a 2020 Energy Snapshot, the Marshall Islands was working with the US Department of Energy to encourage ocean and/or general renewable energy via feed-in tariffs, metering, billing, other incentives, tax credits, or reduction; green procurement preferences; and energy efficiency standards (footnote 99). This plan committed to 75% higher energy efficiency in government buildings and 50% in domestic residences by 2020. It also aimed for a 32% reduction in greenhouse gases. The Government of the Marshall Islands and ADB have been involved in various projects, including "Preparing Clean and Renewable Energy Investments in the Pacific" and "Building Coastal Resilience through Nature-Based and Integrated Solutions."

---

[142] M. Beger et al. 2008. *The State of Coral Reef Ecosystems of the Republic of the Marshall Islands*.

[143] J. R. Hein, W. C. Schwab, and A. S. Davis. 1988. Cobalt and Platinum-Rich Ferromanganese Crusts and Associated Substrate Rocks from the Marshall Islands. *Marine Geology*. 78 (3–4). pp. 255–283.

[144] M. D. Conrad. 2016. Republic of the Marshall Islands Pursuing a Sustainable and Resilient Energy Future. National Renewable Energy Laboratory, US Department of Energy.

Marine Renewable Energy Technology Options Relevant to the Marshall Islands

The first MARES challenge in the Marshall Islands is to select viable, scalable multifunction capabilities from the MRE technologies that exist or are in development at various technology readiness level (TRL) maturities. Table 5 provides an overview of ocean–energy options for the Marshall Islands, presenting estimates for energy potential and costs. It also includes estimates of LCOE and levelized cost of hydrogen (LCOH). The full range of assumptions underpinning these estimates is available at Appendix 2.

This is followed up by initial analysis of each platform's applicability within this particular island setting, building on the more general explanations of the varying technologies that were presented in Chapter 4.

Table 5: The Republic of Marshall Islands' Marine Renewable Energy Options and Indicative Estimates

Technology	Potential [Theoretical] Capacity (GW)	Capital Cost ($ million/MW)	Levelized Cost of Energy ($/kWh)	Levelized Cost of Hydrogen LCOH (based on LCOE) ($/kg)	Reference LCOH by 2030 ($/kg)
Marine Solar	2,290.77–22,907.66[a][b][c]	1.50–1.88	0.094–0.134	7.81–11.16	1.55–2.5[f]
Wave	0.037–0.605[a][b][c][d][e]	2.7–9.1	0.066–0.866	5.46–72.14	Not Available
OTEC	12.83–128.28[a][b][c]	3.00–13.00	0.021–0.091	1.75–7.58	6.79–9.51[g]
Offshore Wind	4,982.03–49,820.27[a][b][c]	3.00–4.00	0.069–0.091	5.71–7.61	3.50–6.14[h][i]
Marine Bioenergy	4.58–45.76[a][b][c]	3.50–4.50	0.040–0.051	3.33–4.28	Not Available
Tidal/Current	67.35–673.49[a][b][c]	3.30–5.60	0.377–1.279	31.39–106.54	Not Available
Salinity Gradient[j]	No data	27.50–35.00	No data	No data	No data

GW = gigawatt, kg = kilogram, kWh = kilowatt-hour, LCOE = levelized cost of energy, LCOH = levelized cost of hydrogen, MW = megawatt, OTEC = ocean thermal energy conversion.

Note: Underpinning assumptions (listed as alphabets in the table) for these calculations are available in Appendix 2.

Sources: Asian Development Bank (ADB). 2014. *Wave Energy Conversion and Ocean Thermal Energy Conversion Potential in Developing Member Countries*. Manila; A. S. Kim and H. J. Kim, eds. 2020. *Ocean Thermal Energy Conversion (OTEC) Past, Present, and Progress*. London; Central Intelligence Agency. *The World Factbook; Global Wind Atlas*; International Renewable Energy Agency. 2022. *Global hydrogen trade to meet the 1.5°C climate goal: Part III – Green hydrogen cost and potential*. Abu Dhabi; N. Dinh. 2022. *Projections of levelized costs of hydrogen (LCOH)*. MARES Report. Manila: ADB; Pacific Islands Ocean Observing System. *Data Services*; S. Banerjee, M. N. Musa, and A. B. Jaafar. 2017. *Economic assessment and prospect of hydrogen generated by OTEC as future fuel. International Journal of Hydrogen Energy*. 42 (1). pp. 26–37; and World Bank. 2022. *A Roadmap for Offshore Wind in the Philippines*. Infographic. 25 April.

Offshore Wind (High Likelihood of Viability in the Marshall Islands)

The average wind speed in the Marshall Islands is 7.09 m/s at 100 m elevation with a corresponding power density of 310.69 watts per square meter (m²).[145] Figure 28 illustrates average hypothetical wind energy potential for the archipelago. The wind speeds are higher in the northernmost islands, while assessment could be worthwhile for Majuro and Ebeye (footnote 96) including utility scale (>500 kW) offshore turbines in their lagoons.[146] However, data is still limited because of less than 20 years of meteorological data being available.[147] Past measurements done by US agencies show average wind speeds of 6–7 m/s.[148] An offshore wind turbine was used as shore

---

[145] Technical University of Denmark. *Global Wind Atlas*.

[146] Government of the Marshall Islands. 2018. *Navigating our Energy Future: Marshall Islands Electricity Roadmap*. Republic of the Marshall Islands Energy Future.

[147] IRENA. 2015. *Renewables Readiness Assessment: Republic of the Marshall Islands*. Abu Dhabi.

[148] Government of the Marshall Islands. 2016. *National Energy Policy and Action Plan 2016*. Majuro: Energy Planning Division of the Ministry of Resources and Development.

protection in Arno Atoll, with the base being grown as an artificial reef using mineral accretion techniques.[149] However, reef-mounted wind turbines may also affect the reef ecosystem (footnote 147). REN21's Renewables Global Status Report states that a 1.5–7.5 MW wind turbine has a capacity factor of 35%–45%.[150]

Figure 28: Wind Power Density Map of the Republic of the Marshall Islands

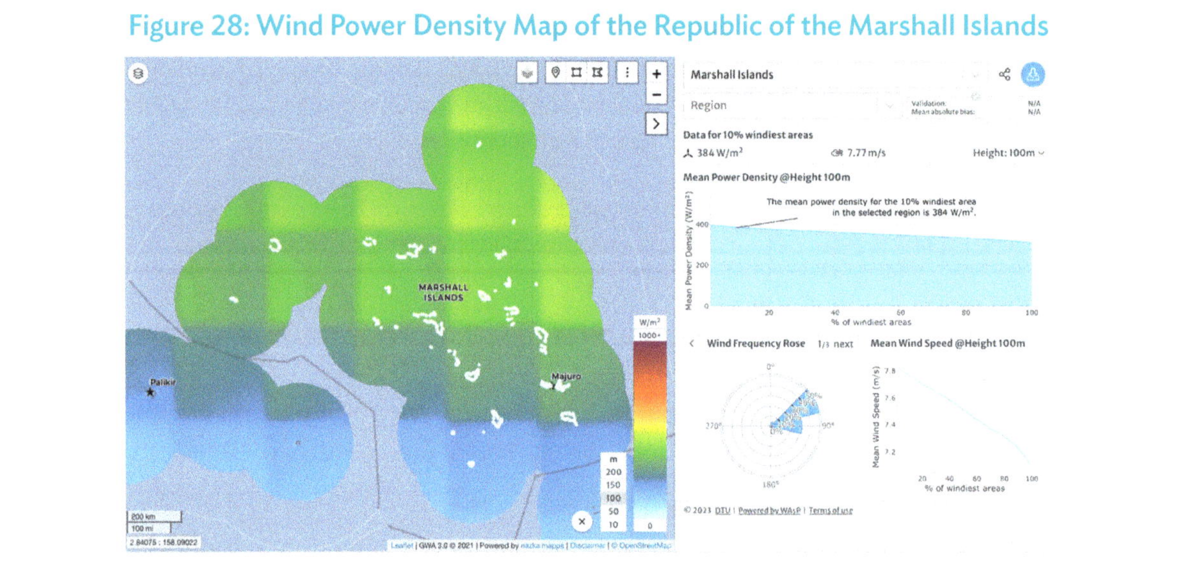

m = meter, m² = square meter, N/A = not applicable, s = second, W = watt.

Source: Map obtained from the Global Wind Atlas 3.0, a free, web-based application developed, owned and operated by the Technical University of Denmark (DTU). The Global Wind Atlas 3.0 is released in partnership with the World Bank Group, utilizing data provided by Vortex, using funding provided by the Energy Sector Management Assistance Program (ESMAP). For additional information: https://globalwindatlas.info.

Marine Solar (High Likelihood of Viability in the Marshall Islands)

Marine solar systems, which may be attached to a fixed structure or floating on the body of water, are perhaps the most promising of all the types of marine renewable resources available to the Marshall Islands. According to studies, the Marshall Islands' solar resource is greatest in the northern islands and least in the middle islands (Figure 29). The average solar potential is at 5 kWh/m² per day.[151] Floating solar photovoltaic (FPV) is being explored, with a pilot in the Majuro lagoon (footnote 146). A pilot project for shore protection using mineral accretion in Majuro Atoll was also commissioned (footnote 149).

Figure 29: Photovoltaic or Solar Energy Resource Potential

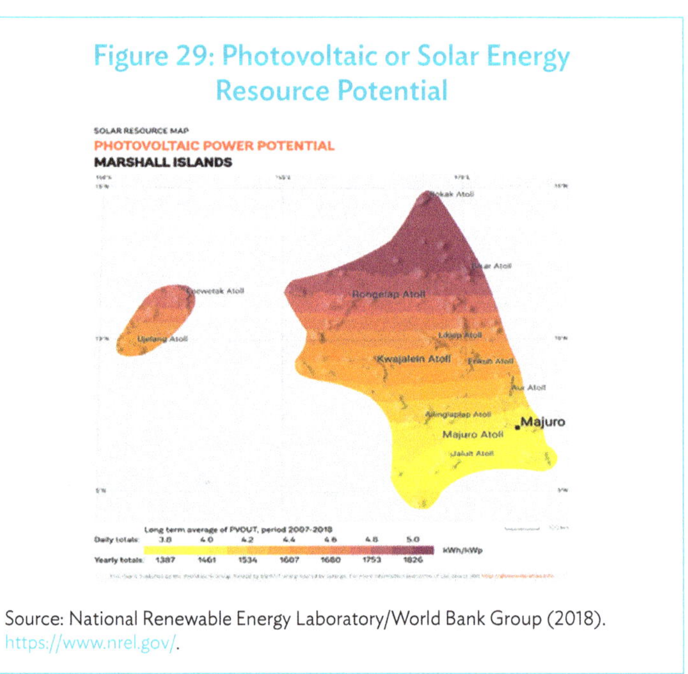

Source: National Renewable Energy Laboratory/World Bank Group (2018). https://www.nrel.gov/.

---

[149] K. Jormelu, E. Hagberg, and T. J. Goreau. 2010. *Shore Protection in the Republic of the Marshall Islands: Pilot Project Report*. Cambridge: Global Coral Reef Alliance.

[150] REN21. 2015. *Renewables 2015 Global Status Report*. Paris.

[151] National Renewable Energy Laboratory. RE Data Explorer.

Ocean Thermal Energy Conversion (Potentially Viable in the Marshall Islands)

The potential for ocean thermal energy conversion (OTEC) within the Marshall Islands has been explored. Estimated annual generation of 1,430 gigawatt-hours (GWh) for a 100 MW plant was predicted by a previous study.[152] But OTEC was still considered to have a low TRL and thus unsuitable for deployment until 2025 (footnote 146). Given the commercial successes of small OTEC facilities in Hawaii and Japan, and the fact that many OTEC systems have multi-user capacity potential for underwater data centers, ocean cooling, mariculture, research centers, and artificial reefs—none of which have been fully explored—they could generate further economies of scale.

Studies made by the National Renewable Energy Laboratory show that the resource potential for OTEC, once plants are established in the Marshall Islands, is 380 GWh per year. Kwajalein Island, which houses a US military base, is considered as the location for a 20 MW OTEC power plant to provide electricity and fresh water to the military facility. But operating OTEC plants also poses multiple operational challenges, aside from the environmental impacts and the associated land requirements.

Financial considerations, and not technical ones, represent the biggest barriers at present to the more widespread adoption of OTEC technology, especially at capacities below 10 MW. The basic electricity generation process system is simple, and both the Hawaii Natural Energy Laboratory and the Okinawa Deep Seawater Research Center have proven that the technology is reliable. A financial guarantee is needed to go beyond small demonstration plants to precommercial prototype units, which will give key operational performance data required to encourage investment in even bigger commercial-scale facilities.

Tidal Flows and Currents (Potentially Viable in the Marshall Islands)

Tidal energy is not mentioned in the National Energy Policy (footnote 148) and is stated as unsuitable in the short term (footnote 146). A tidal-powered shore protection project using mineral accretion was planned but canceled because of manufacturer's delays (footnote149). There may be specific sites or island groups that have potential for tidal energy (e.g., harnessing tidal range and/or currents), but further investigation is needed.

The potential for ocean as well as for tidal current energy remains experimental, unproven globally with relatively few pilot projects and no examples of commercialization. No previous studies have yet been conducted specifically for the Marshall Islands. Theoretically, the capacity is estimated to be commercially unviable, with limited potential, compared to other ocean–energy sources given the comparatively further position of the Marshall Islands between the North Equatorial Current and the Equatorial Counter Currents.

Wave Energy (Potentially Viable in the Marshall Islands)

A 10 kW/m wave energy potential was assessed for Mauro Atoll.[153] It was considered for Gugeegue Island in the 1990s but never developed (footnote 97). A wave-powered shore protection project using mineral accretion was undertaken in Arno Atoll (footnote149). However, there were no well-tested commercial wave energy systems (footnote 147) present at the times of previous studies, and wave energy is mostly deemed unsuitable until 2025 (footnote 146). Nevertheless, the potential for wave energy exploitation is present; and, since the wave energy resource is present in the Marshall Islands, further investigation may lead to possible projects. In the past few

[152]   ADB. 2014. *Wave Energy Conversion and Ocean Thermal Energy Conversion Potential in Developing Member Countries*. Manila.
[153]   A. Lauranceau-Moineau. 2019. *Blue Energy: Renewables in the Pacific Ocean*. Presented during the workshop on Incorporating the Ocean in NDCs. Fiji. 9 May.

years, more innovative efforts and technological innovations have surfaced, several of which have been detailed as possible solutions by the MARES team, especially in combination with other sources.

Salinity Gradient (May Be Unviable in the Marshall Islands)

There is no specific literature for salinity gradient energy for the Marshall Islands. The National Renewable Energy Laboratory and RE Data Explorer also do not provide data for the said resource (footnote 151). The country also does not have large freshwater bodies to provide a salinity gradient resource.

Marine Bioenergy (Potentially Viable in the Marshall Islands)

The Marshall Islands Office of Commerce, Investment and Tourism stated that algae grown in a controlled environment may be a candidate technology to increase the country's renewable energy share.[154] There may be potential associated to mass-scale growth of seagrass meadows, seaweed for consumption, and utilization of algae from extended protected areas or ecosystem rehabilitation.

Roundup of Options Applicable to the Marshall Islands

This review shows that, of the assessed MRE platform types,

- two (offshore wind and marine solar) appear very likely to be viable for deployment in the Marshall Islands;
- four (OTEC, tidal flows and currents, wave, and marine bioenergy) look potentially viable, but more analysis is required; and
- one (salinity gradient) looks likely to be unviable.

This analysis will help to inform the next stages of assessment.

Blue Economy Sector Opportunities in the Marshall Islands

MARES consultants have undertaken initial analysis of the program's key sectors of interest as they relate to the Marshall Islands to understand local context and prepare the ground for potential multifunction projects. Such analysis also helps to develop the essential criteria that will need to be met for MARES projects to succeed. Examples are as follows:

- As a result of overfishing, sea cucumbers have been depleted in many Pacific island countries, and the aquaculture of these animals may represent a very interesting commercial prospect. Culture can take the form of hatchery and release to boost fisheries, with an alternative being growth through to commercial size. It is likely that technical capacity would have to be increased to enable scaling of sea cucumber farming. The Pacific clam, *Asaphis violascens,* has also been looked at for culture in the Marshall Islands. The commercial potential of these and giant clams and fish aquaculture will be considered, alongside other MARES multifunction elements in the Marshall Islands.

---

154   Government of the Marshall Islands. n. d. *Renewable Energy for the Marshall Islands.* Majuro: Office of Commerce, Investment and Tourism.

- On the island of Majuro, coral reefs of the near-shore lagoon adjacent to Delap-Uliga-Darrit are heavily impacted by pollution, runoff, marine debris, and overexploitation (footnote 142). Thus, there could be potential for MARES projects focused on restoration of these reefs; however, that could be a waste of resources unless the issues leading to the negative impacts are mitigated.

- The tourism sector in the Marshall Islands contributes a little under 10% of the country's GDP.[155] An estimated 5.5% of the country's work force, or just over 600 Marshallese, were working in the tourism industry in 2015,[156] but 38% of jobs were in the private sector. Even though locals are being paid more, the Marshall Islands' tourism industry is increasingly relying on foreign workers. The 2015 employer survey reported that it is difficult to fill jobs in the tourism industry—from entry-level jobs like waiters and retail staff, to those requiring advanced qualifications and training.[157] The majority of the staff at high-end resorts, like Beran Island, are imported, especially chefs and trained surf and dive instructors, because skills are not locally available.[158] This implies that any MARES initiative should take capacity development as a central consideration (and not just for ecotourism projects). This consideration was also made directly to project consultants by in-country stakeholders.

- The Marshall Islands' current tourism sector growth strategy is defined within a document called the Strategic Development Plan 2020–2024. The plan proposes a target of 300% growth in room occupancy and 60,000 visitor arrivals by 2024.[159] The Government of the Marshall Islands also supports tourism in its National Strategic Plan 2020–2030, outlining tourism as a significant economic opportunity, which suggests that it has "untapped potential in contributing to income and employment creation."[160] The plan emphasizes the potential of the blue economy and urges greater local involvement and joint venture investments in the sector. This provides useful leverage for potential MARES projects, as their multifunction nature typifies the blue economy in action.

Potential Pilot Sites for MARES Multifunction Projects

The final step in moving from the theoretical to the actual in the Marshall Islands involves considering suitable locations to conduct pilot projects that can prove the success of each technical concept; support the marine, blue, and ocean regenerative symbiosis economy; and provide sufficient evidence for full commercial scalability.

The Marshall Islands consists of about 1,225 islands and 870 reef systems spread over 750,000 square miles of the Central Pacific. For practical purposes of the site identification process, the Marshall Islands was divided into three categories: (i) Majuro and Kwajalein Atolls, where the majority of the population resides; (ii) atolls with a population of less than 2,500 people; and (iii) all remaining uninhabited reefs and atolls. There are two atolls with a population of over 2,500 people (Kwajalein and Majuro).

Following a process of assessment, the following three Marshall Islands sites were selected as being of greatest applicability to the MARES project: Majuro, Rongrong, and Eneko Island (Figure 30).

155 South Pacific Tourism Organisation. Research and Stats Update: June 2021.
156 South Pacific Tourism Organisation. 2019. 2018 Annual Visitor Arrivals Report. Suva.
157 ADB. n.d. The Marshall Islands—A Private Sector Assessment. Unpublished.
158 Pacific Private Sector Development Initiative. 2021. Republic of the Marshall Islands: Pacific Tourism Sector Snapshot. Sydney.
159 Government of the Marshall Islands. 2019. Strategic Tourism Development Plan 2020–2024. Majuro.
160 Government of the Marshall Islands. 2020. National Strategic Plan 2020–2030: Republic of the Marshall Islands. Majuro.

Figure 30: Potential MARES Project Sites in the Marshall Islands

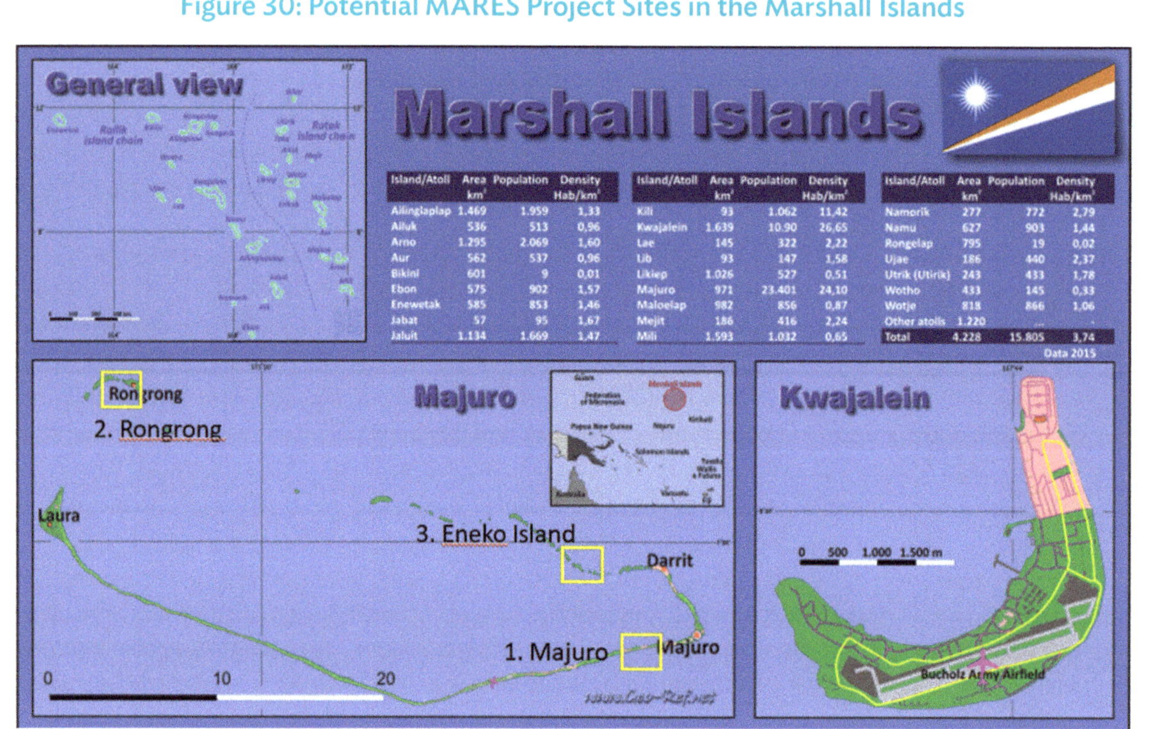

Island/Atoll	Area km²	Population	Density Hab/km²	Island/Atoll	Area km²	Population	Density Hab/km²	Island/Atoll	Area km²	Population	Density Hab/km²
Ailinglaplap	1.469	1.959	1,33	Kili	93	1.062	11,42	Namorik	277	772	2,79
Ailuk	536	513	0,96	Kwajalein	1.639	10.90	26,65	Namu	627	903	1,44
Arno	1.295	2.069	1,60	Lae	145	322	2,22	Rongelap	795	19	0,02
Aur	562	537	0,96	Lib	93	147	1,58	Ujae	186	440	2,37
Bikini	601	9	0,01	Likiep	1.026	527	0,51	Utrik (Utirik)	243	433	1,78
Ebon	575	902	1,57	Majuro	971	23.401	24,10	Wotho	433	145	0,33
Enewetak	585	853	1,46	Maloelap	982	856	0,87	Wotje	818	866	1,06
Jabat	57	95	1,67	Mejit	186	416	2,24	Other atolls	1.220	…	
Jaluit	1.134	1.669	1,47	Mili	1.593	1.032	0,65	Total	4.228	15.805	3,74

MARES = Marine Aquaculture, Reefs, Renewable Energy and Ecotourism for Ecosystem Services.
Source: Geo-Ref.net (2022). http://www.geo-ref.net/.

Majuro

Majuro is one of the Marshall Islands' 29 atolls and home to about half of the entire population as well as the international airport. It covers a land area of 9.2 km². Majuro is the site of most tourism accommodations and commercial infrastructure. There is potential for any MARES energy project to support existing offerings that now rely on the electrical grid.

Rongrong

Rongrong is the largest islet on the Northwest corner of Majuro lagoon. It is close to the international airport and a short boat ride inside the lagoon to the population center of Majuro. There are also safe anchorages on the south side of the island inside the lagoon. There is a high school on the island, and Rongrong is running off the grid exclusively from diesel generators. It is thus far enough to operate off-grid as a case study for all off-grid islands, but close enough to easily resupply with the necessary labor and materials.

Eneko Island

Eneko, on Majuro's northern shore, is 9 km from the international airport and is a short boat ride away. The island offers pristine beaches; clear, shallow waters; and a few underwater relics from World War II. The area is popular with snorkelers because of the colorful fish and coral. It is known locally as the "conservation island," so it offers significant potential for eco- or nature-based tourism initiatives.

Potential MARES Projects in the Selected Marshall Islands Pilot Sites

The kinds of MARES project that could be promoted in the Marshall Islands include a mix of ideas aligned to the specific sites of interest. Suggested projects have been included to show the thinking on the range of criteria that need to be considered when developing a MARES project.

Majuro

- Install a wave-powered living Gabion wall Breakwater (a low-cost shoreline engineering structure) for fuel depot.
- Tidal energy production from flooding and ebbing tides.
- Mighty fly and electric vertical take-off and landing demonstration project for emergency services to outer atolls.

Rongrong Island

- A 30-square mile no-take marine managed zone; fish and coral nursery for lagoon restocking.
- Power the island 100% from hybrid renewable energy as a "fossil-free" zone, as a demonstration project to power 27 off-grid atoll communities with a population from 50 to 2,500.
- Vocational school tied to local high school for hardware software training.
- Near-shore floating wind turbines (2x locations 6 MW each).
- Kite sail turbine demonstration project (1 MW).

Eneko Island

- Standardized, modular, plug-and-play off-grid tourism development demonstration project.
- Starlink Communication package for remote environmental sensors and monitoring, distance learning and medicine, and gig economy employment.

Conclusion

The Marshall Islands appears well-positioned to receive MARES multifunction pilot projects, and there are a number of maturing and scalable innovations that align well to the needs of the islands and their communities.

Offshore wind and marine solar have been identified as being highly likely to be viable in the waters of the Marshall Islands; OTEC, and marine bioenergy are all potentially viable. Salinity gradient is, at this stage, not considered suitable for deployment in the Marshall Islands.

CHAPTER 9

In Focus: Palau

Having explored some of the more promising MRE and blue economy sector options for the Marshall Islands, this chapter undertakes the same review of Palau, in the context of multifunction opportunities in the island.

Marine Renewable Energy in Palau

The Republic of Palau includes eight principal islands and more than 250 smaller ones located roughly 500 miles southeast of the Philippines, in Oceania. The islands of Palau constitute the westernmost part of the Caroline Islands chain. The country includes the World War II battleground of Peleliu and world-famous rock islands. The total land area is 459 km^2 (177 square miles). It has the 42nd largest EEZ of 603,978 km^2 (233,197 square miles).

The Marine Aquaculture, Reefs, Renewable Energy, and Ecotourism for Ecosystem Services (MARES) project team reviewed Palau's national preexisting policies and strategies to prepare the ground for ensuring that emerging MARES project proposals align with these and do not conflict or contradict in any way.

Existing electricity infrastructure and operations are undertaken by the Palau Public Utilities Corporation, with a network extending over 47 linear miles of connections.[161] This offers primarily diesel-based generators, with minor contributions by renewable solar energy for the remoter Southwest Islands, Tobi, and certain main users on Koror and/or Babeldaob Islands.

For marine or ocean–energy sources, no current or projected demonstration sites exist for offshore wind, wave, tidal, salinity gradient, current, floating solar voltaic, and other forms of marine or ocean energy or alternative fuel sources. Palau also previously aimed for a 30% reduction in energy consumption by 2020. The country's 2009 Energy Policy indicates that the energy sector aims to reach 20% renewable energy by 2020 and improve energy efficiency by 2030. As part of this process, Palau was supported through a 2009–2013 project to improve energy efficiency subsidies and audits for households, by the drafting or implementation of the National Energy Efficiency Action Plan, and by other technical support.

In parallel, the United Nations Development Programme and the International Union for the Conservation of Nature investigated solar energy during that time, excluding a Palau International Airport solar energy project. A Palau government 2012 assessment indicated high solar energy capacity at over 5.5 kWh/m^2 per day.[162] Wave energy was calculated as less effective at 10–15 kW/m. It considered that offshore wind energy, hydroelectric, tidal, and current power still needed an effective technical and commercial assessment, while ocean thermal energy conversion (OTEC) might have to wait for the future to generate sufficient energy efficiency and economies of scale.

---

[161] Renewable Energy and Energy Efficiency Partnership. Republic of Palau (2012).
[162] UN Climate Technology Centre and Network. 2013. Republic of Palau (2012).

Although not yet deployed, Palau has previously contemplated OTEC as far back as 2002, as it produces zero emissions, requires little space, and is inexpensive, with perpetual energy consistently generated.[163] Its technical feasibility has been suggested at several locations able to offer at least 24 MW of capacity, including desalinated potable water and support for aquaculture. After a prefeasibility study from Japan's Saga University confirmed this, the option was not favored due to lack of funding and other minor issues. The initial pilot project would have generated 3 MW of OTEC energy, the spray flash method for desalinated water at 1,000 cubic meters (m^3) per day and 1 normal cubic meter per hour of hydrogen per cycle.

Palau's National Energy Policy does not specifically mention ocean or MRE, but these sources can help support its vision and ultimate objective of "a reliable and resilient energy sector delivering Palau sustainable, low emissions energy services, by maximizing cost-effective energy efficiency and renewable energy resources and conservation of energy while safeguarding our environment."[164] It also provides an opportunity for hydrogen and other alternative fuels as a basis for reducing dependency on fossil fuels, resilience, and adaptation to climate change and improved energy security. The policy also creates a role for improved education and awareness and an Energy Administration, absorbing the previous Energy Office, to guide the actual implementation, investment, and operation. This is echoed further in the 2016 Palau Energy Act, which requires reporting related information under the Pacific Islands Framework for Action on Climate Change, with commitments to renewable energy installation and energy efficiency compliance via a monitoring and evaluation framework and annual reports.[165] It further deals with specific responsibilities to maintain and operate renewable solar energy programs and other forms of energy, administer any related grants and funding, while formulating and implementing any regulations. Both the government transport fleet and marine vessels used in tourism and ferries are expected to comply with certain efficiency standards to reduce contributions to pollution and climate change.

A series of 2013 reports by the International Renewable Energy Agency on Pacific renewable energy also probed OTEC, offshore wave and wind, floating solarvoltaic, hydropower, and biomass for Palau, although without providing any additional support or confirming research.[166] It estimated that this could solve the use of over 10,033,249 liters of fuel for the Malakal power plant, 2.6 megaliters for fuel, and 43.4 megaliters for shipping fuel in 2013. It confirmed that many detailed assessments including offshore wind and wave energy have not yet been conducted. The government Capitol Building, main hospital, airport, and Departments of Education and Public Works, along with the National Archives, invested in 546.26 kWh of solar energy off-grid solutions.

The vision of the Palau Climate Change Policy is "happy, healthy, sustainable and resilient Palauan communities in a changing world." From 2020 onward, ADB's Pacific Renewable Energy Investment Facility has been investigating floating solar power projects across 11 Pacific nations, including Palau and the Marshall Islands.[167] The project aims to be completed in phases starting with Kiribati, Tonga, and Tuvalu. The project committed an initial $2,000,000 for project preparation, feasibility consultancy studies, gender mainstreaming, and technical capacity building.

Palau has an ambitious target of 45% renewable energy by 2025 and 100% renewable by 2050. The Government of Palau is planning to add an independent solar power plant operator to contribute a further 13 MW. The government is seeking $100,000,000 in potential investment to achieve its renewable energy goals under

[163] G. Decherong. 2002. *The Project for Ocean Thermal Energy Conversion (OTEC) and its Multi-purpose Utilization in the Republic of Palau.* Prepared for the forum on Desalination using Renewable Energy. Palau. 15–16 October.
[164] Palau Energy Policy Development Working Group. 2012. *Palau National Energy Policy.* Koror.
[165] Government of Palau. 2016. *Palau Energy Act.* Ngerulmud.
[166] IRENA. 2013. *Pacific Lighthouses: Renewable Energy Opportunities and Challenges in the Pacific Islands Region—Palau.* Abu Dhabi.
[167] ADB. 2020. *Technical Assistance for Preparing Floating Solar Plus Projects under the Pacific Renewable Energy Investment Facility.* Manila.

the Disaster Resilient Clean Energy Financing initiative of ADB and reaching out to the private sector.[168] It offers around a 30 MW grid, versus demand at 12.5 MW average use. Solar renewable energy only occupies <7% of the total or 2.7 MW. This is 93% supplied by the Palau Public Utilities Corporation. Additionally, in 2019, diesel-fueled generators cost over $19,000,000 in imports, aside from contributing significantly to high total fossil fuel emissions and emissions per capita. Palau's Energy Roadmap and Policy have estimated high availability of solar energy for up to 100% of potential requirements at a rate of 5.5–5.9 kW/m^2 per day. Wind energy has been estimated at an average rate of 2.5 m/s.

The first stage of the process reviewed options for replacing extant carbonaceous-fueled electricity grids with sustainable, 24/7, consistent renewable energy supply, either on the national grid and/or local grids for more remote locations. It considered alternative fuel sources such as hydrogen, ammonia, and bioenergy to reduce the need for fossil fuels, committing to decarbonization, climate-neutral economy in alignment with the 2015 Paris Agreement on Climate Change.

Marine Renewable Energy Technology Options Relevant to Palau

As with the Marshall Islands, the MARES challenge focuses on how to select viable, scalable, multifunction capabilities from the many MRE technologies that exist or are in development at various TRL maturities. Table 6 provides an overview of ocean–energy options for Palau, presenting estimates for energy potential and costs. It also includes estimates of LCOE and LCOH. The full range of assumptions underpinning these estimates is available at Appendix 3.

This is followed by an initial analysis of each platform's applicability within this particular island setting, building on the more general explanations of the varying technologies that were presented in Chapter 4.

Table 6: Palau's Marine Renewable Energy Options and Indicative Estimates

| | | | | | Levelized Cost of Hydrogen | |
Technology	Range	Potential [Theoretical] Capacity (GW)	Capital Cost ($ million/MW)	Levelized Cost of Energy ($/kWh)	LCOH (based on LCOE) ($/kg)	Reference LCOH by 2030 ($/kg)
Marine Solar	Low	17.26[a][b][c]	1.50	0.1339	11.16	2.5[i]
	High	678.68[a][b][c]	1.88	0.0938	7.81	1.55[i]
Wave	Low	0.1519[a][c][d]	2.70	0.8657	72.14	Not Available
	High	0.3076[a][b][c][©]	9.10	0.0656	5.46	Not Available
OTEC	Low	0.10[a][b][c]	3.00	0.0890	7.42	9.51[j]
	High	3.80 [a][b][c]	13.00	0.0205	1.71	6.79[j]
Offshore Wind	Low	13.84[a][b][c]	3.00	0.1522	12.68	6.14[k][l]
	High	544.30[a][b][c]	4.00	0.1142	9.51	3.50[k][l]
Marine Bioenergy	Low	0.04[a][b][c]	3.50	0.0514	4.28	Not Available
	High	1.51[a][b][c]	4.50	0.0400	3.33	Not Available

continued on next page

---

[168]   ADB. 2020. *Proposed Administration of Grant to the Republic of Palau for the Disaster Resilient Clean Energy Financing.* Manila.

Table 6 continued

Technology	Range	Potential [Theoretical] Capacity (GW)	Capital Cost ($ million/MW)	Levelized Cost of Energy ($/kWh)	Levelized Cost of Hydrogen	
					LCOH (based on LCOE) ($/kg)	Reference LCOH by 2030 ($/kg)
Tidal/Current	Low	5.07[a][b][c]	3.30	0.8524	71.03	Not Available
	High	50.74[a][b][c]	5.60	0.2511	20.93	Not Available
Salinity Gradient[m]	Low	No data	27.50	No data	No data	No data
	High	No data	35.00	No data	No data	No data

GW = gigawatt, kg = kilogram, kWh = kilowatt-hour, LCOE = levelized cost of energy, LCOH = levelized cost of hydrogen, MW = megawatt, OTEC = ocean thermal energy conversion.

Note: Underpinning assumptions (listed as alphabets in the table) for these calculations are available in Appendix 3.

Sources: Asian Development Bank (ADB). 2014. *Wave Energy Conversion and Ocean Thermal Energy Conversion Potential in Developing Member Countries*. Manila; A. S. Kim and H. J. Kim, eds. 2020. *Ocean Thermal Energy Conversion (OTEC) Past, Present, and Progress*. London; Central Intelligence Agency. *The World Factbook; Global Wind Atlas*; International Renewable Energy Agency. 2022. *Global hydrogen trade to meet the 1.5°C climate goal: Part III – Green hydrogen cost and potential*. Abu Dhabi; N. Dinh. 2022. *Projections of levelized costs of hydrogen (LCOH)*. MARES Report. Manila: ADB; Pacific Islands Ocean Observing System. *Data Services*; S. Banerjee, M. N. Musa, and A. B. Jaafar. 2017. *Economic assessment and prospect of hydrogen generated by OTEC as future fuel*. *International Journal of Hydrogen Energy*. 42 (1). pp. 26–37; and World Bank. 2022. *A Roadmap for Offshore Wind in the Philippines*. Infographic. 25 April.

Marine Solar (High Likelihood of Viability in Palau)

The average solar potential is at 5–6 kWh/m^2 per day (footnote 151). A separate study by the Palau government (2013) puts the estimate of solar energy capacity at over 5.5 kWh/m^2 per day. The Government of Palau is active in seeking renewable energy development based on their policies on energy and sustainability. Palau has a successful on-grid rooftop solar program implemented. As of December 2020, there is an approved project financed by the Asian Clean Energy Fund under the Clean Energy Financing Partnership Facility amounting to $2 million for preparing floating solar, as well as projects under the Pacific Renewable Energy Investment Facility. The proposed transaction technical assistance facility will assess the potential and feasibility, develop a road map, and build institutional capacity to deploy floating PV projects in these 11 small Pacific island countries. There is also a 13 MW floating solar initiative proposed by the Republic of Palau. Using only 1% of the EEZ of Palau already represents an estimated 600 gigawatt-peak of marine and/or offshore solar PV potentially installed.

Wave Energy (Potentially Viable in Palau)

A 2 kW/m wave energy potential was assessed for Palau (footnote 152). On the other hand, another source mentions that the wave energy resource is estimated to be in the 10–15 kW/m range in Palau (footnote 152), making it likely viable. However, no commercial wave energy technologies have yet been tested specifically for Palau, making it unlikely to be used in the near future (footnote 166). Nevertheless, the potential for wave energy exploitation is present; and, since the wave energy resource is present in Palau, further investigation may lead to possible projects. In the past few years, more innovative efforts and technological innovations have surfaced, especially in combination with other sources.

Ocean Thermal Energy Conversion (Potentially Viable in Palau)

A 100 MW OTEC plant was estimated to generate 1,460 GWh per year (footnote 153). A more specific study estimated a 30 MW OTEC potential across 7 identified sites, with a pilot project on the first site starting at 3 MW, with expected significant freshwater and hydrogen production (footnote 168). The spray flash method is used for desalinated water at 1,000 m³ per day and 1 normal cubic meter per hour of hydrogen per cycle. A 2004 study stated that OTEC technology can replace up to 25,000–30,000 barrels of oil per hectare.[169]

Table 7 shows the potential of OTEC in various locations in Palau, covering the whole country. Most of the locations are in the eastern coast of the country. The steep slope on the eastern side provides the temperature difference needed to operate OTEC technology.

OTEC, however, is becoming far more viable in Palau than previous studies may suggest, enhancing its possible application and relevance as among the more technically feasible, efficient, cost, and eco-efficient based on criteria such as location, reliable supply of energy, cost, and return on investment. Unlike solar energy, it also offers potential to be reliable nocturnally.

Table 7: Assessment of Potential OTEC Locations in Palau

Potential OTEC Site	Size of OTEC Plant
Airai	3 MW x 2 + 4 MW x 1 = 10 MW
Melkeok	3 MW x 2 + 4 MW x 1 = 10 MW
Angaur	2 MW x 1 = 2 MW
Ngaraard	2 MW x 1 = 2 MW
Ngarchelong	2 MW x 1 = 2 MW
Ngiwal	2 MW x 1 = 2 MW
Peleliu	2 MW x 1 = 2 MW
Total	**30 MW**

MW = megawatt, OTEC = ocean thermal energy conversion.
Source: G. Decherong. 2002. *The Project for Ocean Thermal Energy Conversion (OTEC) and its Multi-purpose Utilization in the Republic of Palau.* Prepared for the forum on Desalination using Renewable Energy. Palau. 15–16 October.

Financial considerations, and not technical ones, currently present the biggest barriers to the more widespread adoption of OTEC technology especially at capacities below 10 MW. The basic electricity generation process system is simple, and both the Hawaii Natural Energy Laboratory and the Okinawa Deep Seawater Research Center has proven that the technology is reliable. A financial guarantee is needed to move beyond small demonstration plants to precommercial prototype units, which will give key operational performance data required to encourage investment in even larger commercial-scale facilities.

Tidal Flows and Currents (Potentially Viable in Palau)

Palau is listed among the developing nations in the Pacific Ocean that could receive significant benefits from tidal energy. In addition, efficient technology to capture tidal energy is asserted to be already cost effective.[170] Certain lagoons, channels, and areas could have possible exploitable tidal flows and currents, but further investigation is needed. According to the US Insular Areas Energy Assessment Report, the most promising of opportunities for Palau include doing a basic survey of lagoon and open sea passages for possible tidal flow generation.

---

[169]   A. Binger. 2004. *Potential and Future Prospects for Ocean Thermal Energy Conversion (OTEC) in Small Islands Developing States (SIDS).*
[170]   A. Takhar. 2010. *Tidal Energy Overview.* A PowerPoint presentation.

Offshore Wind (Potentially Viable in Palau)

The average wind speed in the Republic of Palau is 4.7 m/s at 100 m elevation, with a corresponding power density of 114.57 watts/m^2 (footnote 145). Figure 31 illustrates the wind power density of the country. Palau's Energy Roadmap and Policy has estimated wind energy at an average rate of 2.5 m/s. Based on climate data measurements, it is unlikely to be economically useful for power generation (footnote 166). Detailed wind resource assessment should be completed if further consideration is to be given to wind power.

Figure 31: Wind Power Density Map of the Republic of Palau

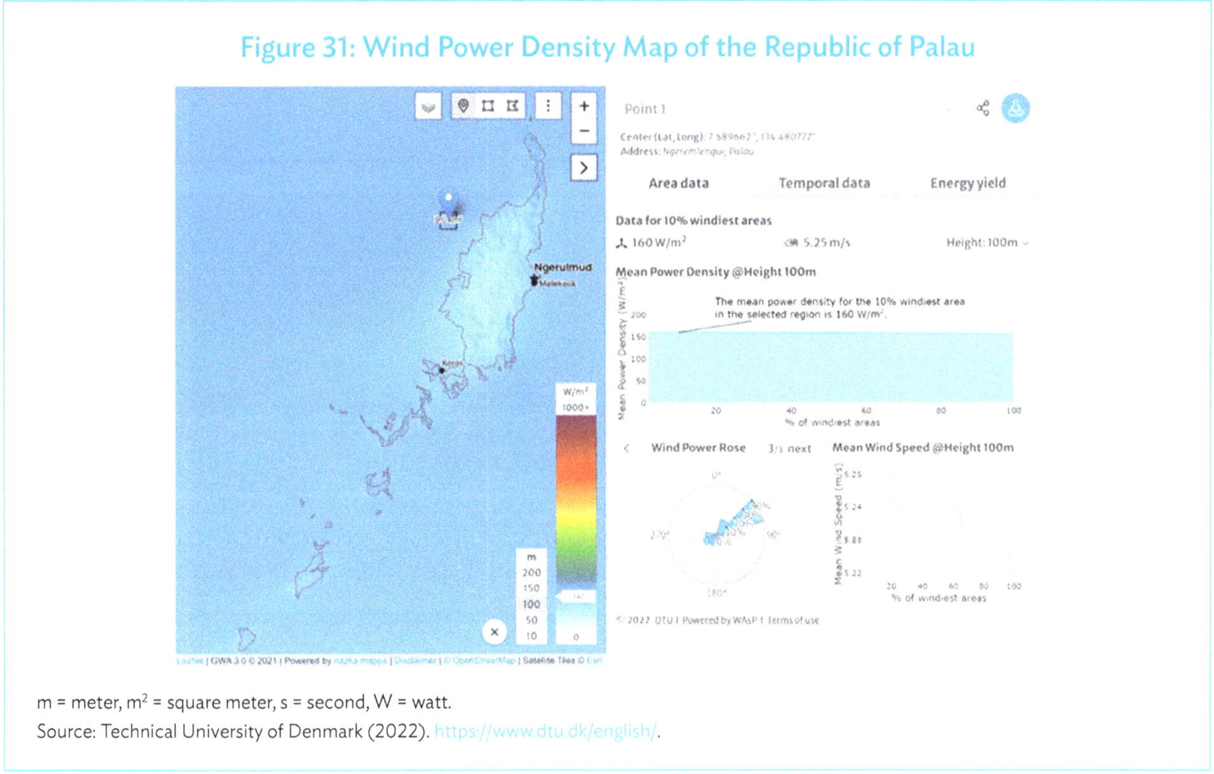

m = meter, m^2 = square meter, s = second, W = watt.
Source: Technical University of Denmark (2022). https://www.dtu.dk/english/.

Salinity Gradient (May Be Unviable in Palau)

There is no specific literature for salinity gradient energy potential and its use for Palau. RE Data Explorer also does not provide data for the said resource (footnote 151). In addition, no large rivers or freshwater channels interface with the sea in the country. Further investigation and studies are requisites for the country to exploit energy from salinity gradient.

Marine Bioenergy (May Be Unviable in Palau)

No studies have been undertaken measuring the marine bioenergy resource potential as well as its use in Palau. There may be potential associated to mass-scale growth of kelp forests, seagrass meadows, seaweed, utilization of *Sargassum* from extended protected areas or ecosystem restoration, and seaweed.

Roundup of Options Applicable to Palau

The MARES project team's initial review shows that, of the assessed MRE platform types,

- one (marine solar) appears very likely to be viable for deployment in Palau;
- four (wave, OTEC, offshore wind, and tidal flows and currents) look potentially viable, but more analysis is required; and
- two (marine bioenergy and salinity gradient) look likely to be unviable.

As with the Marshall Islands review, this analysis will help to inform the next stages of assessment.

Blue Economy Sector Opportunities in Palau

MARES consultants have undertaken initial analysis of the program's key sectors of interest as they relate to Palau, to understand local context and prepare the ground for potential multifunction projects. Such analysis also helps to develop the essential criteria that will need to be met for MARES projects to succeed. Examples are as follows:

- While macroalgal culture may be feasible in Palau, the siting of such facilities has to be carefully considered with respect to potential impacts on coral reef ecosystems. There are potential macroalgal species that could form the basis of a seaweed cultivation industry in Palau. Pickering (2006) reviews some species that could be valuable as food or as sources for the biotechnology or food industry.[171]

- Giant clam (*Tridacna* spp) is the species that has generated most interest and activity in Palau for aquaculture. The first giant clam culture in the world was pioneered in Palau. This is a very high-value species for the Asian market and could also be sold into the local tourist market as a local delicacy. The benefit of bivalve culture is that there is no need to feed the organisms (unlike many aquaculture finfish species that are carnivorous) so environmental footprint and impacts are lower. Bivalves can also improve water quality by taking up particulate organic material. Technical capacity probably needs to be increased to enable scaling of bivalve culture, and this must not be overlooked in any MARES development. There is an underutilized aquaculture center in Palau at the old Micronesian Mariculture Development Centre (now Palau Mariculture Demonstration Centre).

- Palau has a well-established sustainable tourism strategy. In December 2016, the Palau Responsible Tourism Policy Framework 2017–2021 was published, aiming to transition Palau's tourism model from low-end, mass market to a niche, high-value model aligned with Palau's environmental and cultural aspirations. To achieve the shift, the policy prioritizes (i) tourism development and management across the environment, labor, agriculture, and fisheries sectors; (ii) managing visitor economy by attracting high-value niche markets, such as free independent travelers; (iii) developing high-quality standards to brand Palau as a pristine paradise in the global tourism sector; (iv) educating both visitors and the local community to ensure a rich and rewarding experience for all; (v) ensuring economic contribution and revenue retention in the local economy; and (vi) promoting community-driven tourism development.[172] The implementation requires strong partnership and a shared vision across government, the tourism sector, and communities. All these developments speak very well to the MARES approach, so the proffering of more multifunction project ideas, which align well to existing policy aspirations, should be well received.

---

171   T. Pickering. 2006. Advances in Seaweed Aquaculture among Pacific Island Countries. *Journal of Applied Phycology*. 18. pp. 227–234.

172   Government of Palau. *Palau Responsible Tourism Policy Framework 2017–2021*. Koror: Ministry of Natural Resources, Environment and Tourism.

- Cultivation of reef corals and other species can be used both to supply the ornamental fish trade as well as act as a source for reef restoration. This can provide a valuable industry within Palau, but also act as a tourist attraction. Selective breeding (assisted evolution) can also improve the ability of coral to survive in warming waters.

Potential Pilot Sites for MARES Multifunction Projects

The final step involves considering suitable locations to conduct pilot projects that can prove the success of each technical concept; support the marine, blue, and ocean regenerative symbiosis economy; and provide sufficient evidence for full commercial scalability.

More than 250 islands and islets make up the Palau archipelago, which spans along a 150 km north and south trending arc in the western Pacific. Its center is roughly located near 7° north latitude, roughly 650 km north of Jaya on the island of New Guinea, and near 134° east latitude, some 900 km east of Mindanao, Philippines.

Babeldaop, primarily a volcanic island with a 363 km^2 area that makes up more than three-quarters of Palau's total land mass, dominates the geologically diverse archipelago. The other 90 km^2 are occupied by three volcanic islands, two atolls, and numerous uplifted coralline limestone islands. These limestone islands known as the "rock islands" are primarily located in the archipelago center between Oreor and Beliliou islands. The limestone islands of Beliliou and Angaur, otherwise known as platform islands, are not considered rock islands for various physiographic and historical reasons. Southwest Islands are several small islands, including Sonsorol and Hatohobei (Tobi), which are located some distance southwest of Angaure.

After a process of assessment by MARES project consultants, the following three Palau Islands sites were selected as being of greatest applicability to the MARES project: Melekeok State, Koror State, and Peleliu Island (Figure 32):

Melekeok State

Melekeok is on the main island of Babelbaob, which is the second largest and least developed land mass in all of Micronesia. It is a target area for special economic zone development and largely covers large open flat areas denuded of forest.

This area was of great interest to MARES project consultants. The existence of an industrial quality pier, the deep channel to the open sea, strong tidal currents, and large waves for wave energy equipment testing all make this area an ideal MRE location.

Melekeok is also the most popular site for surfing. According to TripAdvisor, "For many years this was one of Palau's best kept secrets mostly due to the fact that travel to the area was difficult and there were only a handful of local surfers (mostly Palauans who picked up the sport living overseas)." Although surf tourism in Melekeok is currently left only to the most

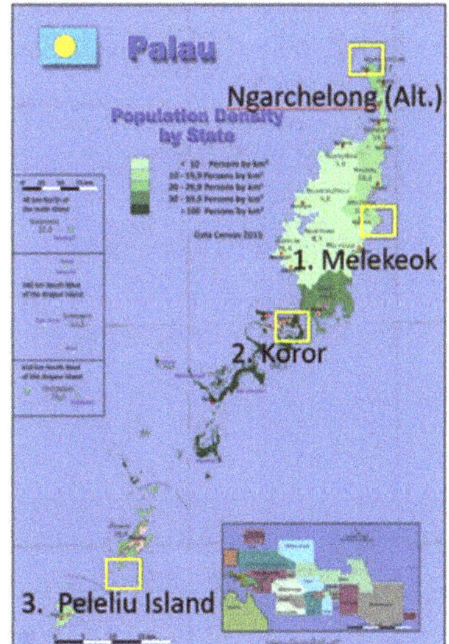
Figure 32: Potential MARES Project Sites in Palau

MARES = Marine Aquaculture, Reefs, Renewable Energy and Ecotourism for Ecosystem Services.
Source: Geo-Ref.net (2022). http://www.geo-ref.net/.

adventurous of travelers, it could be further expanded to become a main tourist attraction by the creation of net-zero surf lodges that are powered by renewable energy.

MARES consultants currently consider this area as being able to support a number of renewable energy and ecotourism projects, potentially operating in close proximity.

Koror State

Located in the center of Palau's central business, tourism, and population centers, Koror is the closest connection to access to the main electric grid and the location where most of Palau's energy is being used.

A previous 2001 study was undertaken for a pilot project for the government by Japan's Saga University and Xeneysis. Though not publicly available, its findings indicated the technical feasibility of marine energy along east coast-based islands, including Koror, to address all electricity demand requirements.

For this area, consultants have identified at least three potential projects that cover tidal energy, FPV, aquaculture, and ecotourism.

Peleliu Island

Peleliu is one of Palau's 16 states and is located 26 miles south of Koror. The third most populated state in Palau, Peleliu Island in Southern Palau has 13 km^2 of landmass surrounded by mangroves, shallow seagrass-covered sand flats, and coral reefs. The island has a population of approximately 500 residents in 135 households and is powered almost exclusively from aging diesel-fueled generators. The island has an international airport, three small resorts, and world-class scuba diving; is a breeding ground for cornerstone marine species; and has a strong deep cold ocean current, suitable for seawater air-conditioning (SWAC), 50 m from the entrance of a World War II-era unused manufactured harbor.

From a tourism lens, Peleliu is well-known as a historical World War II site where one of the largest battles in the Pacific took place between the US and Japan. The main tourism activities that are currently being offered are heritage, adventure, and nature-based tourism.

Potential MARES Projects in the Selected Pilot Sites in Palau

The kinds of MARES multifunction projects that could emerge include a mix of ideas aligned to the specific sites of interest. These ideas have been suggested by the project consultants, in part to develop thinking on the range of criteria that need to be considered when developing a MARES initiative.

Melekeok State

For this area, MARES consultants envision as many as five renewable energy and ecotourism projects operating nearby. Phase I of the development could involve a small-scale onshore OTEC/SWAC research facility for MRE production at the end of an existing government pier at Ngermecheluch.

The pilot project could validate assumptions for the economic feasibility of a 5,000 km^2 facility at the end of the existing government pier for a 3 MW land-based OTEC, SWAC, and 1 MW slow water tidal turbine facility at the edge of a deep-water channel leading out to the open ocean and to a drop off to 1,000 m depth within 1 km distance of the pier.

The existence of an industrial quality pier, the deep channel to the open sea, strong tidal currents, and large waves for ecotourism promotion and wave energy equipment testing all make this area an ideal ocean renewable energy (ORE) location.

In Phase II, it could be possible to build the proposed OTEC, SWAC, tidal, and/or ORE facility and then, using available government land near the capital building, install 20 MW of agrivoltaics and 8 MW of wind energy to meet the majority of all of Palau's energy needs from a hilltop renewable energy facility tied into the main grid. The entire location becomes a tourist attraction and enough freshly grown herbs, leafy greens, and vegetables are grown under the solar panels via aquaponics and stored in the SWAC cold storage facility, as needed, to meet the majority domestic demand.

For enhanced additionality, an ecotourism initiative could also be considered. Although surf tourism in Melekeok is currently left only to the most adventurous of travelers, it could be further expanded to become a main tourist attraction by the creation of net-zero surf lodges that are powered by renewable energy.

Koror State

For this area, MARES consultants have identified at least three indicative projects that cover tidal energy, FPV, aquaculture, and ecotourism. These explorative projects are as follows:

KB Bridge Tidal Energy. The first opportunity is to install a test turbine for slow-water energy harvesting in the 2 km-long and 30-meter-deep channel under the Japan-Palau Friendship bridge (KB Bridge). In the pilot phase, at a depth to avoid collision with vessels navigating the channel, a single 100 kw slow-water tidal turbine could be placed to study its output and any impacts on surrounding marine life. If proven to be reliable and economical, more units could then be added to scale up to 5 MW peak capacity. Additionally, this could lead to the development of a new potential tourism asset that would require relatively low financial investment. A mini-interpretation center could be created at the base of the bridge on either the Koror or Airai side to educate visitors and residents on tidal energy and other forms of MRE. As the base of the bridge is publicly owned lands, negotiation would only require the involvement of governors of each state and, of course, authorization from the Palau Public Land Authority. There is an opportunity to make this structure a landmark iconic feature of the island by commissioning an artistic and innovative monument.

Landmark Marina: MARES administrative office, visitors center, and departure dock for Peleliu Island. Depending on the scope and scale of MARES activities in Palau, it may warrant establishing a showcase for the projects to create tours, attract investors, use for sales, conduct public relations, recruit and train local employees, and facilitate logistics to the various project sites. This location is strategically located next to the Palau Aquarium and Palau International Center for Coral Reef Research for collaborations.

Nico Bay Floating Solar. Land is at more of a premium price the closer you get to the center of Palau's commercial hub. But the entire area is connected via a maze of ocean waterways and small coves, making it ideal for FPV. Some of these back bays are sheltered from wind and waves and only several hundred meters from a potential grid tie connection. The MARES team has identified several potential locations where FPV would not create a hazard to navigation and, if combined with marine aquaculture, could even have a positive effect on surrounding seawater quality in the stagnant back bays that see relatively little water movement.

Peleliu Island

The primary objective of one potential project within this area could be to create a "fossil-fuel-free zone" and convert 100% of the residences and businesses on Peleliu Island to run exclusively on renewable energy. The project could start with 2 hectares of solar power to produce 1 MW peak plus additional complimentary vertical axis wind turbine and ORE. That would help to determine the optimal mix of low-cost dependable solar baseload and other renewable energy to reduce energy storage requirements from batteries to supply 100% uptime for the electricity to meet all near future demand from residents and future "power-to-x" MARES businesses in an industrial park incubator environment. Other ideas include

- upgrading to electric vessels and vehicles on the island, including swappable batteries and a daily electric- or hydrogen-powered ferry service to and from the main commercial center in Koror;
- establishing a for-profit aquaculture incubator facility powered by MRE to highlight various mariculture technologies in the 4-hectare Kambek harbor and surrounding reefs;
- converting the existing three resorts to become carbon negative for optional day tours from Koror and long-stay guest looking to have a guilt-free carbon-negative vacation; and
- developing scalable containerized hydroponic and agrivoltaic food production facilities.

These very early-stage outline ideas—alongside other proposals from interested companies and other stakeholders in the Marshall Islands and beyond—will be assessed further by MARES project consultants, using the process outlined in Chapter 10.

Conclusion

Marine solar energy has been identified as being highly likely to be viable in the waters of Palau—wave energy, OTEC, tidal flows or currents, and offshore wind are all potentially viable. Marine bioenergy and salinity gradient are, at this stage, not considered suitable for deployment in Palau.

Melekeok State, Koror State, and Peleliu Island have all been identified as having the right range of characteristics to support and benefit from MARES pilot projects. Ngarchelong State has been identified as the next best candidate location.

CHAPTER 10

A Principles-Led Approach

The Marine Aquaculture, Reefs, Renewable Energy, and Ecotourism for Ecosystem Services (MARES) multifunction approach involves bringing together multiple innovations, each of which may be complex in its own right. To support the Marshall Islands, Palau, and other island states to maximize the potential of their blue economies, the MARES project set out to find capabilities, innovations, technologies, or infrastructure that can be combined with one or more other technologies and/or infrastructure to create a multifunction capability with a viable, scalable business model that enables island states to maximize the socioeconomic benefits of their blue economies and large EEZ resources to create and equitably distribute greater prosperity for all.

This penultimate chapter of the handbook sheds light on some of the steps devised and used by the multidisciplinary team of MARES consultants to achieve that clarity of approach to articulate a process that can be replicated in and for other nations looking to adopt the MARES approach. A range of marine renewable energy (MRE) sources collectively forms the starting point for a wider engagement across the use of energy. Figure 33 provides some structure to the new ocean–energy economy—harnessing renewable energy from the ocean, converting that energy into other fuels, boosting mariculture, promoting marine ecotourism, and rehabilitating coral reefs as a source of food and coastal protection.

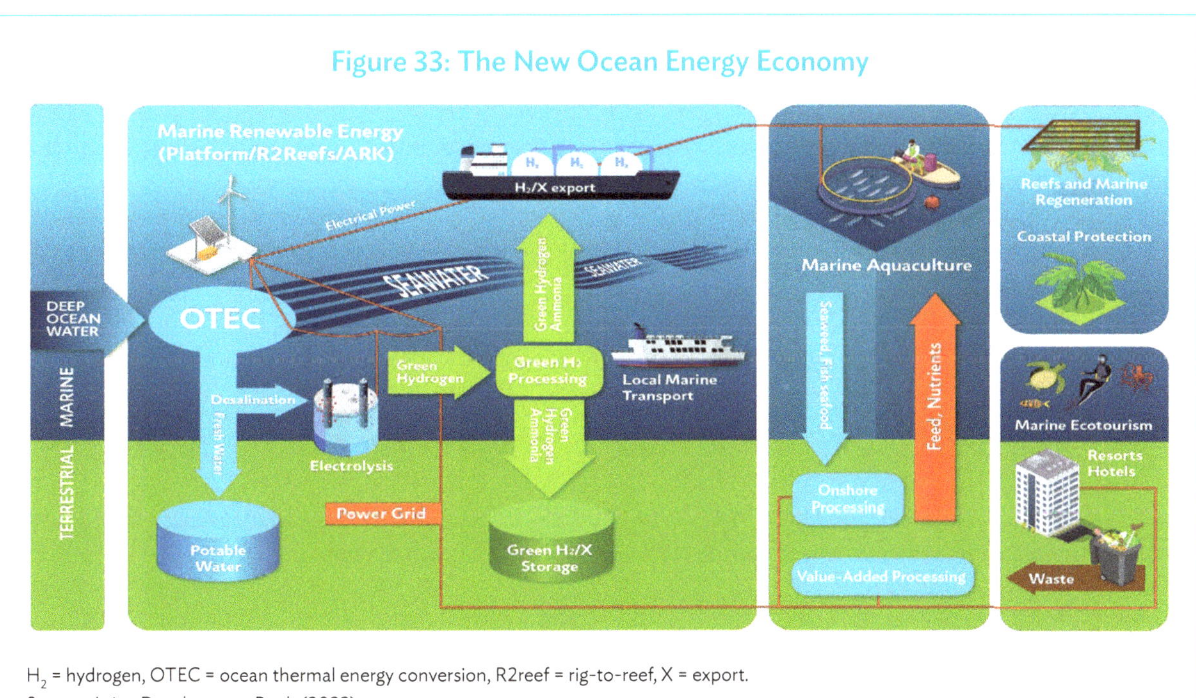

Figure 33: The New Ocean Energy Economy

H$_2$ = hydrogen, OTEC = ocean thermal energy conversion, R2reef = rig-to-reef, X = export.
Source: Asian Development Bank (2023).

Interactions with shipping, ports, fisheries, and value-added business using seawater are not included in the scope of ADB's MARES project (TA 6619-REG). This new ocean–energy economy forms a subset of the wider blue economy.

The National Indicative MARES Plan Process

The MARES multifunction concept is novel, but an approach that has been developed with replicability in mind. To achieve project objectives, it was necessary to develop a selection and screening process to define the right solutions for the intended beneficiary nations, and to establish a set of decision-making tools that could be used elsewhere.

This approach, named the National Indicative MARES Plan process, developed steps to identify multifunction marine and maritime projects that are potentially able to radically transform the economy, environment, and society; to become more resilient and future-proofed against changing circumstances; and to commit toward an ultimately truly long-term regenerative, sustainable, optimum future. The process developed by a team of international experts followed these steps, each of which is covered in greater detail below:

- Exploring and assessing MRE options.
- Achieving a mutual consensus on technology and solution type.
- Selecting potential investable projects based on agreed criteria.
- Identifying pilot sites within the target beneficiary nations potentially suitable for projects.
- Undertaking social and environmental screening to ensure they are applicable to the selected small island developing state (SIDS), given resources, constraints, risks, and priorities.
- The first three steps were explained earlier. The final two stages are considered in this chapter.

Criteria for Candidate Technologies and Projects

As part of the MARES initiative, a number of candidate investable projects were reviewed and presented in the proceedings report of the TA 6619 MARES High-Level Investor Forum 2023, a compendium to this handbook. These projects are, by their very nature, multifunction and, therefore, align to a range of different success metrics. To shortlist potential candidate projects, the MARES team arrived at a set of five crosscutting assessment criteria, namely that projects

- Support and align to existing national aims and objectives:
    - They address national government socioeconomic and ecological issues (e.g., the impact of climate change, protecting and regenerating the marine environment, a just transition to renewable energy, jobs, national wealth, etc.).
    - They are easily understood by government and related stakeholders to enable policy development, regulation, financial incentives, and long-term economic sustainability.
- Are multifunction, technically sound, and scalable:
    - They can be combined with one or more other technologies and/or infrastructure to create a multifunction capability, perhaps in a phased approach over a long timeline. This could include, for example, a coastal or offshore renewable energy installation that creates green marine fuels and supports marine aquaculture while protecting and regenerating an endangered coral

reef, simultaneously providing coastal protection to low-lying island states on the frontline of climate change.

- They are at an advanced technology readiness level (TRL 7–9) and have reviewed previous experiences and recorded failures in operating businesses, traditional energy, etc., drawing on both in-country and international expertise.

- They have proven capability to date or potential capability in 3–5 years.

- Are financially viable:

- They are at an appropriate commercial readiness level (used by investors to assess market readiness) and have developed a robust business model that indicates how they will recover costs and be profitable in order to attract capital from private and corporate investors and international banks.

- They align, where appropriate, to potential commercial markets that may exist for new forms of energy and fuel, and any related customers or usages.

- They contribute to measurable impact investment and sustainable blue finance indicators, such as the United Nations Environment Programme Finance Initiative Blue Economy Principles.

- Are sensitive to local conditions and stakeholders:

- They are locally suitable to the Marshall Islands or Palau and other potential pilot sites that identify physical sites and constraints for such locations and necessary enabling social and environmental aspects.

- They are socially and politically acceptable—welcomed by the public as making a positive contribution to their nation and their communities and by existing blue economy sector operators for their crosscutting benefits.

- They align to existing marine and maritime projects in ways that are mutually supportive.

- Are future-proofed:

- They actively support the reduction of fossil fuels and a decarbonized economy, consider developing climate change conditions, future-proof against disasters, consider likely future technology changes, and adapt sensibly to a COVID-19 recovery process and the mitigation of other risks.

- They aim to enhance local resilience and reduce socioeconomic and ecological vulnerability.

- They consider the potential costs of inaction and alternative scenarios to strengthen the business case.

Criteria for Pilot Sites

With assessment criteria in place for the potential projects, similar decision tools were required to help select the locations where such projects could be deployed. The MARES team agreed on the following areas of assessment:

- **Location.** The site chosen will need to have the required benefits of close access to seawater, power, and enough population so that locals can be hired, as well as provide dwellings for foreign workers. The site will need to complement all these factors to even be considered suitable.

- **Water quality.** Water parameters are a critical issue with aquaculture. Sites selected in target countries must keep this factor a priority on the selection process. The best seawater will generally be found in the ocean as opposed to lagoonal waters as the nutrients are much lower and other parameters such as turbidity, temperature, and salinity are much more stable. This can be essential with many aquaculture species.

- **Ocean access.** One of the greatest expenses in ocean-related projects can be the intake system, depending on the location and natural forces expected on that intake system. Intake depth, ocean energy, intake length, and the required volume of water are key indicators in developing the design and estimates for the structure.

- **Public perception.** A significant factor in potential success is the perception of the public. If done with culture and heritage in mind from the onset, this can be turned into an overall positive driving force to help the project be successful. It is critical to have active engagement with the community to be challenge-ready for doubters and people with legitimate concerns. Historical preference and culture must be made a priority from the onset (including traditional culture, memorials, and the remnants of World War II).

- **Community support.** This is a forever changing dynamic in the process. It should be considered in the design of the project to allow for change of governors, presidents, and local politicians and opinions. This can go from positive support to negative very quickly, and care should be taken in recruiting the representatives, ensuring they are truly in a neutral position that is focused on the project.

- **Local knowledge.** Hiring the right locals will be a critical part of the project's success. There is an incredible amount of local knowledge related to history, biology, and logistics that are all relevant to the design, build, and operation of any facility in remote locations.

- **Export potential.** The potential for export is incredibly high when producing the right product. Going forward, the flights from these locations will improve and should give access to Asian as well as US and European markets.

Assessing Potential Social Impacts and Risks

With technical, economic, and market feasibility assessed, and the suitable locations considered, the final set of considerations are more ethical in nature. Detailed social and environmental screening needs to take place to ensure that the projects do not have unintended consequences that could harm the local environment or communities.

The potential risks associated with various types of MRE options have been assessed through analysis of available literature, as presented in Chapter 5. These will also be subject to the ADB Safeguards Policy (2009). It is important to stress that such potential risks, once identified, can be minimized with appropriate measures. The developers and regulatory authorities should consider the dependence of local populations upon natural resources for their livelihood, while encouraging renewable energy development. Mandatory environmental impact assessment studies for all renewable energy projects will need to be undertaken in a rigorous and impartial manner, ensuring that social safeguards are in place, giving greater powers to local communities, providing employment to residents, and ensuring electricity access to the people who live in vicinity of renewable energy projects.

The following summary of key lessons from the literature review on impacts of offshore wind developments and mitigation measures apply to all MRE projects:

- The success of offshore wind developments depends on effectively managing stakeholder and public concerns; opposition from public and stakeholder may cause the delay or cancelation of projects. It is also crucial to involve stakeholders (particularly local communities) early and often throughout the process. Developing a clear process to maintain this engagement, including hiring community liaisons, may be useful.

- To reduce the spread of misinformation and increase trust in the process, it will be important to maintain regular, effective, and honest communication about the project between the developer and supporting government agencies.

- The conduct of marine spatial planning (or some other open and public planning process) to inform site selection can lessen controversy and conflict that may arise later in the process.

- Those who are living near the coast understandably have strong attachment to the place that can make visual impacts strong drivers during implementation processes.

- It may not be suitable to treat indigenous people as just another stakeholder because they have a unique political status and may be affected by specific laws pertaining to consultation and potential impacts on them. Indigenous people have connections to the proposed wind project sites, i.e., both ocean and land areas, and thus have a unique stake in project outcomes.

- A variety of issues concerning fishermen may need to be addressed, such as the loss of important fishing grounds, crowding in other fishing grounds because of displacement, effects of projects on the health of marine resource populations and their habitats, effects on transit zones, encroachment of wind activities on working-waterfront spaces, and loss of fishing industry jobs or income linked to the other impacts.

- Fishermen may have innovative and contextually appropriate ideas about how the packages should be designed. It may be critical to include them early in the discussions about community benefits packages or other ways to address impacts to their industry.

- Be as open as possible when having dialogues regarding community benefits packages as unexpected stakeholders may be interested in participating in such discussions.

- Environmental effects are one of the highest public concerns about offshore wind. To gain public acceptance, it would be necessary to be transparent about environmental impacts and appropriate design to lessen or minimize those impacts.

4

GOING FORWARD

CHAPTER 11

Recap of the MARES Approach

This handbook has presented an overview of the MARES concept, including

- Briefly reviewing some of the challenges facing the world's seas and oceans and highlighting how the blue economy represents a new, holistic way to maximize marine and maritime economic potential, at the same time as embedding environmental and social safeguards.
- Explaining the aims and objectives of the MARES project and providing contextual detail on its key areas of investigation: mariculture, renewable energy, reef rehabilitation, and ecotourism.
- Highlighting the potential for encouraging greater synergistic collaboration and integration between new initiatives in these sectors.
- Exploring how such opportunities may successfully align in the Marshall Islands and Palau.
- Presenting a new set of decision-making processes that the MARES team of experts has devised to select a shortlist of potential multifunction marine projects and to encourage all interested parties to use these processes to design potential projects in their areas.

Opportunity for Multifunction Marine Projects

The opportunity clearly exists for coastal and island states to maximize the potential of their vast EEZs by encouraging multifunction marine projects that combine several capabilities holistically. Such activities will enable nations to meet their climate commitment while achieving a regenerative marine environment coupled with a just transition to a true local blue economy.

The blue economy concept is no longer at the theoretical stage, as governments around the world are separately and collectively progressing key elements of sector practice along shared principles and constantly refining overarching strategy. More and more, blue economies are being robustly quantified, through innovative environmental and satellite accounting practices. Broader understanding is developing of the huge value of ocean activities as a driver of innovation, economic growth, and prosperity, and in creating jobs.

There is growing clarity within new blue governance structures that economic development in the ocean must go hand in hand with carefully planned and managed regenerative activities. Coordination and integration are the watchwords of such an approach. Both also sit at the core of the MARES concept, which can be seen as a logical tactical extension of the blue economy at an operational level.

A global review of MRE capabilities relevant to coastal and island states has set out which may be most applicable in various regions. The examination of conditions in the Marshall Islands and Palau has not only identified sites and ideas ripe for development but has also produced a replicable set of criteria to guide such assessments.

These criteria, which can in time be used to explore and assess the viability of a broader range of ocean innovations and in different locations, include

- the alignment with existing national policies, regulations, aims, and objectives;
- technical maturity and scalability;
- financial and/or commercial viability;
- sensitivity to local conditions and stakeholder needs; and
- robustness in the face of future scenarios driven by climate change, COVID-19 and/or other external shocks, and other potential risks such as ocean acidification.

There is much more to do: drive the development and adoption of holistic blue economy strategies, improve investment ecosystems, accelerate private finance, and support multifunction marine and maritime projects that exemplify the new opportunities made possible by the blue economy.

However, these tasks can be tackled with much greater confidence in light of two of the key strategic elements presented in this handbook:

- That the conditions for the scaling of MRE are improving significantly, as evidenced in the rapidity of technological advancement and policy and financial support. The need for reliable MRE underpins the potential for any hydrogen economy activity as well as storage, electrification, and other energy use cases.
- That smaller coastal and island states can very much now be a part of this narrative, as the development of marine green hydrogen now enables their massive EEZs to be viewed with huge promise for their economies, their population, and the world at large, as exportable MRE will offer multiple benefits to climate mitigation.

This promises cyclical benefits—investment in MRE in previously unsupported marine estates could provide the economies of scale necessary for significant cost reductions in all parts of the sector, which in turn would make broader adoption a reality.

Conclusion—The Technology and Business Models Are Already Operating

Hundreds of potential innovations have been investigated by the MARES team of experts to assess their viability, applicability, and scalability. Far from being unproven and expensive, many of the technologies competed with their fossil fuel counterparts when transportation, storage, and losses were considered.

As the MARES technical assistance project was carried out during the COVID-19 pandemic, the team was required to adapt to new conditions. It was not possible to carry out workshops in person. A data room was created to host content.[173] A total of 25 webinars were held, explaining the content to the 2,000 people who attended. When loaded online at the data room, over 22,000 separate views of the webinars were reported.

---

[173]   ADB Knowledge Events. *ADB Data Room: Marine Aquaculture, Reefs, Renewable Energy, and Ecotourism for Ecosystem Services.*

A call for projects was issued in July 2022 and received 50 submissions. These submissions were assessed and shortlisted to nine business plans. The participants went through workshops and individual coaching sessions to refine their business models and get ready to present. During this process, two teams combined, and one other team changed their business model based on feedback during coaching.

The eight shortlisted and coached teams presented at a High-Level Investor Forum in Kuala Lumpur in February 2023. This event also engaged with the thinking, opportunities, and challenges presented in this handbook, and highlighted further areas for progress. Appendix 1 contains an extract of the Proceedings of the High-Level Investor Forum.

The most powerful theme of the High-Level Investor Forum is that the technology and business models are already operating. What is missing is an understanding of the opportunities—a kind of sea blindness. That will be addressed in future work.

Extract of the Proceedings of the Workshop on Business Development for MARES Shortlisted Projects and High-Level Investor Forum on the New Ocean Energy Economy

Overview

On 6 and 7 February 2023, the Asian Development Bank (ADB) and the Centre for Indonesia-Malaysia-Thailand Growth Triangle (CIMT-GT) hosted a High-Level Investor Forum (HLIF) in Kuala Lumpur with the theme of the "New Ocean Energy Economy." Over 130 participants discussed marine renewable energy and its use to regenerate ocean ecosystems. The Minister of Economy, Malaysia, Yang Berhormat Tuan Mohd Rafizi Bin Ramli made a keynote speech while representatives from ADB, CIMT-GT, Palau, and the Philippines gave their remarks.

Forum participants. About 130 participants and resource persons from the government, private sector, academia, financing institutions, and civil society organizations gathered for the High-Level Investor Forum on the New Ocean–Energy Economy (photo by Indonesia–Malaysia–Thailand Growth Triangle [IMT-GT] Joint Business Council).

ADB's technical assistance (TA) project, Marine Aquaculture, Reefs, Renewable Energy, and Ecotourism for Ecosystem Services (MARES) (TA 6619-REG), is a knowledge and support project that aims to facilitate future investment in sustainable ocean economy development. The TA has two main activities: (i) assessment of marine resource commercialization prospects (including energy, seafood, and tourism) and identification of potential investment projects in selected developing member countries (DMCs); and (ii) stakeholder engagement and knowledge management on mechanisms to facilitate large-scale investments and to accelerate financing of selected projects.

Specifically, MARES was designed to complete the following tasks:

- Assessment of policy and regulatory framework in target ADB's DMCs;
- Assessment of existing examples of MARES solutions and provision of case studies, where available;
- Identification of potential MARES projects for initial screening, and selection of one project proposal that best meets the objectives of MARES;
- Collation of the project solutions with social and environmental safeguards and economic screening into an ADB knowledge product or handbook;
- Screening of potential transactions with social and environmental safeguards and economic screening into project pipeline with recommendations; and
- Hosting and presenting a High-Level Investor Forum on MARES opportunities.

At the HLIF, representatives from government, infrastructure, and energy developers; investors; blue economy businesses; and community groups gathered to discuss how to realize a "New Ocean Energy Economy." After the speeches and remarks, there were three plenary presentations, with each followed by a moderated panel discussion that allowed the selected participants to delve deeper into the issues that were the focus of each plenary. Prior to the event, a workshop was held on 6 February 2023 to assist the eight shortlisted projects in developing their project pitches through the feedback provided by project facilitators and investors.

A survey was conducted after the workshop to get the views and comments of the participants.

Key Takeaways

The Blue Economy Landscape

- After decades of development, **the blue economy is taking off,** with many new projects coming on stream, which are aligned to strategic developments and investments.
- **Ocean blindness is everywhere.** There are still many senior staff at large development organizations and other institutions who have not appreciated how important the ocean is as a carbon sink, and how a blue economy with renewable ocean projects will be key to stopping the negative impacts of climate change.
- As Jacques Cousteau said, "It is difficult to love what you don't know," **and many people in positions of influence do not appreciate how important the ocean is** in allowing us to breathe and survive.

The Role of Marine Renewable Energy

- **The technology is here; it is just not evenly distributed,** especially among large ocean states. Technologies such as ocean thermal energy conversion (OTEC) are maturing rapidly. The key issue is simply how to scale.
- **Green hydrogen is being created and sold on international markets, but has been slow to develop** in large ocean states (also known as small island developing states).
- Recent events have increased the pricing of carbon-based fuels, making renewable energy more cost effective, and creating **a good time to invest in blue renewable energy**.

The MARES Approach

- **Marine renewable energy can be the catalyst for many other profitable** businesses (multisector, multifunctional, regenerative), especially aquaculture and coral reefs.
- **Integrated marine spatial planning** that involves local cultures and communities is needed for successful MARES-type projects.

What Else Needs to Happen

- **Action leads to attraction.** There is a need to move toward active deployment of profitable projects as soon as possible to demonstrate key blue economy, multifunction, MARES principles if we want to seriously take on climate change.
- **There is a gap between project developers and investors.** ADB can play a key role in helping to match groups of projects with investors and guiding MARES projects to successful funding.
- **More investment forums—online and in person—will help to connect investors with investable projects of different sizes.**
- **ADB has pledged $5 billion via the Healthy Oceans Action Plan (2019–2024)** to support ocean projects but **needs to speed up implementation to reach this goal**.
- **Regional cooperation and integration** can help ADB to fund MARES projects, and **ADB can help to train and facilitate project promoters** to refine their pitches for the various types of investors.

MARES HLIF in Numbers

- 130+ participants
- 28 experts
- 5 developing member countries
- 4 plenaries
- 18 countries
- 8 project pitches from 6 countries selected from 50 submissions

Program

Workshop and Final Judging of Shortlisted MARES Projects
Jasmin Room, Impiana Hotel
Monday, 6 February 2023

11:00 a.m. –12:00 nn	**REGISTRATION**
12:00 nn – 1:30 a.m.	**LUNCH AND NETWORKING**
1:30 p.m. - 2:30 p.m.	**PRESENTATION OF SHORTLISTED PROJECTS**

- Pacific Ocean Explorers (POE), Tom Bowling (Palau)
- Savusavu Blue Town Model, Judy Cakacaka, Bright Tide (Fiji)
- Decom2Green, Robert Young, Pan Ocean Aquaculture (Thailand)
- Deep Water Intake Infrastructure powered by MW-scale OTEC, Benjamin Martin, Japan Industrial Academic Consortium including Institute of Ocean Energy Saga U. (IOES) and Xenesys Inc. (Palau)
- Republic of the Marshall Islands Climate Proof Fuel Storage, Dan Brian Millisson (Republic of the Marshall Islands)
- Blue-Cooking, Tasmiat Rahman, University of Southampton (Bangladesh)
- Ongedaol Nature Resort Palau, Irene Olkeriil, Poshtel   (Palau)
- Subic Blue, QRS Aqua Inc. and Ccell Ltd., Gregor Hodgson (Philippines)

2:30 p.m. – 4:00 p.m	**WORKSHOP: MENTORING AND COACHING SESSION**

Project developers will be grouped into three to get one-on-one coaching sessions to improve business plans and investment pitches.

Jonathan Turner, Director, NLA International Ltd

4:00 p.m. – 4:15 p.m	**AFTERNOON BREAK**
3:15 p.m. – 4:45 p.m.	**PLENARY: SHARING OF INSIGHTS**

Jonathan Turner, Director, NLA International Ltd

4:45 p.m. – 5:00 p.m.	**FINAL REMARKS IN PREPARATION FOR THE HLIF**

Stephen Peters, Senior Energy Specialist (Waste to Energy), Energy Sector Group, Sustainable Development and Climate Change Department, ADB

High-Level Investor Forum on The New Ocean Energy Economy
Kuala Lumpur Convention Centre, Kuala Lumpur Malaysia
7 February 2023

8:30 a.m. – 9:30 a.m.	Registration
9:30 a.m. -9:35 a.m.	Playback of IMT-GT subregional cooperation and Development Partner video
9:35 a.m.– 10:30 a.m.	**OPENING PLENARY**

Opening Remarks

Alfredo Perdiguero, Director, Regional Cooperation and Operations Coordination Division, Southeast Asia Department, ADB

Welcoming Remarks

Firdaus Dahlan, Director, Centre for IMT-GT Subregional Cooperation

Introductory Remarks

- *Eden Uchel, Director, Palau Energy and Water Administration*
- *Romeo M. Montenegro, Assistant Secretary and Deputy Executive Director, Mindanao Development Authority (MinDA), Philippines*

Keynote Speech

The Hon. Rafizi Ramli, Minister of Economy, Malaysia

Moderator: Teymoor Nabili, The Signal Singapore

10:30 a.m. – 10:45 a.m.	**Morning Break**

10:45 a.m.– 1:00 p.m.   PLENARY 1: The New Ocean Energy Economy

Opening Remarks

Netty Muharni, Assistant Deputy Minister for Regional and Sub Regional Economic Cooperation, Coordinating Ministry for Economic Affairs of Indonesia (recorded video)

Introduction to The New Ocean Energy Economy

Nick Lambert, Founder, NLA International Ltd

Panel Discussion on Integrating the New Ocean Energy Economy to Global Energy Needs

- *James Doldolia, Program Coordinator, MinDA Power Development Program*
- *Kee-Yung Nam, Principal Energy Economist, Energy Sector Group from the Sector Advisory Cluster (SDSC-ENE), ADB*
- *Michael Abundo, CEO, OceanPixel Pte Ltd*
- *Wan Sayuti Wan Hussin, Head (Strategy and Policy), Corporate Sustainability, Petronas Sustainability Unit, Petronas BHD*
- *Moderator: Cindy Cisneros-Tiangco, Principal Energy Specialist, PAEN, ADB*

Presentation of Eight Shortlisted MARES Projects

- *Andy Hamflett, Co-Founder and Director, NLA International Ltd*

1:00 p.m. – 2:30 p.m.   Lunch

2:30 p.m. – 4:00 p.m.   PLENARY 2: Using Energy from the Ocean

Opening Remarks

Thuttai Keeratipongpaiboon, Director of International Strategy and Coordination Division, Office of the National Economic and Social Development Council (NESDC), Thailand

Announcement of the Winning Project

Jonathan Turner, Director, NLA International Ltd

Integration of the Hydrogen Economy and Marine Renewable Energy

Ines Marques, Director for the Green Hydrogen Development Plan, The Green Hydrogen Organisation with recorded message from World Bank, UNIDO, and the Green Hydrogen Organisation

Panel Discussion on Integrating Micro-, Small-, and Medium-Sized Enterprises (MSME) into New Ocean Energy Economy

- *Dan Millison, Consultant, SDSC-ENE, ADB*
- *Irene Olkeriil, Founder, EmeralDreams Services, LLC*
- *Salma Khoo, Jaringan Ekologi Dan Iklim*
- *Sarah Dole, Founder, Invena*
- *Scott Countryman, Executive Director, The Coral Triangle Conservancy*

Moderator: James Ellsmoor, Founder and Director, Island Innovation LLC

4:00 p.m. – 4:15 p.m.   Afternoon Break

4:15 p.m.– 5:30 p.m.   PLENARY 3: Responsibility Using Marine Spaces

Opening Remarks

Andreas Dipi Patria, Head of Communication Bureau, Coordinating Ministry of Maritime and Investment Affairs Republic of Indonesia

Panel Discussion on Financing the New Ocean Energy and Blue Economies

- *Alix Burrell, Principal Investment Specialist, Private Sector Operations Department, ADB*
- *Joan Fulton, COO, Ocean Assets*
- *Len George, Principal Energy Specialist, PAEN, ADB*
- *Michael Dethlefsen, Chief, Innovative Finance and International Financial Institutions Division, UNIDO*
- *Vincent Choy, Asia Representative, Delphos*

Moderator: Ghislain De Valon, Lead, Blue SEA Finance Hub, ADB

5:30 p.m. – 5:45 p.m.   CLOSING PLENARY

Summation of Progress

Steve Peters

Closing remarks

Priyantha Wijayatunga, Chief of Energy Sector Group, SDCC, ADB

Day 1: Workshop and Final Judging of Shortlisted MARES Projects

On the day prior to the full forum proceedings, a workshop was conducted to further prepare the representatives from the eight shortlisted competition projects.

The half-day program of activities was designed and led by NLA International Ltd. The workshop had two main objectives:

- To provide attendees with an introduction to ocean financing that they could use to consider future funding options; and
- To provide feedback, coaching, and mentoring from project facilitators and investors to help attendees sharpen their project pitches.

Introduction to ocean financing. Rear Admiral (retired) Nick Lambert, founder, NLA International Ltd presented the status of the blue economy and the various ocean financing options (photo by IMT-GT Joint Business Council).

Introduction to Ocean Financing

After welcoming delegates to the session, the participants were presented with a headline overview of the status of the blue economy and provided with details of various ocean financing options.

Definitions and Government Action

A range of definitions of the blue economy were shared, with consistent crosscutting themes highlighted. Selected developments in governmental approaches to promoting the blue economy were also used to provide context; these included the following:

- The launch of the Sustainable Blue Economy Strategy of the city of Umm Al Quwain, which is the capital and largest city of the Emirate of Umm Al Quwain in the United Arab Emirates. The strategy balances economic growth with environmental stability, aiming to double gross domestic product by 2031 and for the blue economy to contribute 40% of that total. A net-zero emissions target has also been established for 2031, by which time a total of 20% of Umm Al Quwain is set to be dedicated to nature reserves. All these actions combine under the headline are aimed of transforming Umm Al Quwain into the "capital of the blue economy."
- The awareness-raising and stakeholder engagement elements of Canada's blue economy strategy. In December 2020, the Canadian Fisheries and Oceans Minister announced that an online consultation process would take place in the following months to allow "provinces, territories, indigenous peoples and others" to contribute ideas to its emerging blue economy strategy.
- Belize's national debt swap for ocean conservation and blue economy development, which aims to lead to a reduction in the country's national debt of BZ$0.5billion ($0.25 billion). Following the model set by Seychelles, the first country to undertake an oceanic debt for nature swap, Belize is partnering with The Nature Conservancy to undertake the transaction and in the process create a Marine Conservation Endowment Account, supporting marine conservation in perpetuity.

- Indonesia's Blue Financing Strategy, which prioritizes the principles of sustainability and helps investors channel their funding through a range of financial instruments into sectors that will have the greatest impact in terms of sustainability. While focusing on supporting the archipelagic state to grow its economy faster, the strategy also makes clear that unlocking investments for the development of the blue economy is a climate action priority for archipelagic and island states.

Selected Financing Mechanisms

Attendees were presented with high-level details of several development bank initiatives related to the blue economy. These included

- The World Bank's PROBLUE Programme,
- ADB's $5 billion Action Plan for Healthy Oceans and Sustainable Blue Economies,
- The European Investment Bank's Clean and Sustainable Ocean Programme, and
- The Ocean Innovation Challenge run by the United Nations Development Programme.

Key similarities and differences in approach between these programs were highlighted, to underscore the need to be flexible with story-telling tools to facilitate the widest possible interest.

Exercise: Mapping with and Alignment to the UNEP Blue Economy Finance Initiative

A group exercise was led to test the participants' potential to tell the story of the impact they would like to achieve with their innovations in flexible ways. Project leaders were provided with details of the United Nations Environment Programme (UNEP) Sustainable Blue Economy Finance Initiative. They were asked to align the 14 program principles (e.g., protective, compliant, inclusive, solution-driven, etc.) to their own project and show if, where, and how they supported the program's aims.

Mentoring and Coaching Sessions

Project participants were supported to practice telling their stories and receive group and one-on-one feedback to improve engagement tactics and investment pitches.

Day 2: High-Level Investor Forum on the New Ocean–Energy Economy

Opening Plenary

Opening Remarks by Alfredo Perdiguero, Director, Regional Cooperation and Operations Coordination Division, Southeast Asia Department, Asian Development Bank

Alfredo Perdiguero commended CIMT-GT for being known as a leader in green development, and for working on the issue of developing green cities all over this region for many years. He was delighted to know that CIMT-GT is going into the area of the blue economy.

At ADB, blue financing is also an emerging area with increased interest from investors, financial institutions, and issuers globally. It offers tremendous opportunities to help safeguard our access to clean water, protect underwater environments, and invest in a sustainable water economy. Understanding the risks and the opportunities (for blue financing) is a critical first step in formulating and adopting well-calibrated and well-harmonized policies. ADB is now trying to approach the multiple crisis facing the oceans or "poly-crisis" from a multisector perspective. This poly-crisis needs informed solutions and requires the best available knowledge and a supportive policy environment.

The Asia and Pacific region is highly vulnerable to the impacts of climate change. Promoting a green and blue recovery is important to saving the planet. Toward this end, ADB has committed to deliver $100 billion in climate finance by 2030 to speed up the momentum to tackle climate change risk, decarbonize the region, and leverage innovative financing to support green and blue projects across the region. To fulfill this role, ADB launched the Action Plan for Healthy Oceans and Sustainable Blue Economies. This initiative also included the commitment to provide technical assistance and investments of $5 billion for 2019 to 2024. ADB is committed to supporting the blue economy through the Blue Southeast Asia Finance Hub and the Blue Pacific Finance Hub.

Opening remarks. Alfredo Perdiguero, director, Regional Cooperation and Operations Coordination Division, Southeast Asia Department, highlighted ADB's support to blue economy (photo by IMT-GT Joint Business Council).

"Here at ADB, blue financing is an emerging area in climate finance with increased interest from investors, financial institutions, and issuers globally. It offers tremendous opportunities to help safeguard our access to clean water, protect underwater environments, and invest in a sustainable water economy."

Welcome Remarks by Firdaus Dahlan, Director, Center for Indonesia, Malaysia, Thailand - Growth Triangle

Firdaus Dahlan noted that the forum's topic, Innovative Alternative Energy Solutions Using Ocean Energy to Benefit Our Economies, was quite relevant to the Indonesia–Malaysia–Thailand (IMT) region as it has abundant maritime resources. Commenting that, "Environmental sustainability should become the paradigm for our current and future development agenda," he noted that IMT has already formally adopted the Green-Blue Circular Economic framework in its latest Implementation Blueprint 2022–2026 to ensure the sustainability of the regional and subregional environment. The IMT has allocated a budget of $16.9 million for investment in activities related to the green cities projects, which will reduce greenhouse gas emissions by 6 million tons of carbon dioxide per year. By exploring the new ocean–energy economy, Dahlan said that IMT will align its development goals and new alternative ocean–energy initiatives. Noting that the subregion is connected to the Andaman Sea, the Indian Ocean, the Gulf of Thailand, and the Malacca Straits, he said, "If we can leverage this geography by promoting alternative energy and ecotourism then we can unlock new income sources and ensure the environmental sustainability for the subregion. I hope we can identify investment opportunities today."

"In our current Implementation Blueprint, Indonesia-Malaysia-Thailand – Growth Triangle adopts the green, blue, and circular economy approach to ensure the sustainability of our subregional environment. Moreover, this approach not only shows our commitment to align our development strategy with the global environmental agenda but also becomes our strategy to tap new markets and investment opportunities."

Welcome remarks. Firdaus Dahlan, director Center for Indonesia, Malaysia, Thailand - Growth Triangle (CIMT-GT) reiterated that IMT-GT has already adopted the green, blue, and circular economy approach to ensure environmental sustainability (photo by IMT-GT Joint Business Council).

Introductory Remarks by Eden Uchel, Director, Palau Energy and Water Administration, Ministry of Finance, Palau

Eden Uchel stated that as Palau is at the forefront of climate change, the nation's "cumulative voices must engender actions that will not only improve socioeconomic development, but also ensure our common future, and pave a path toward financing and moving ideas to implementation." She noted that Palau fully subscribes to Sustainable Development Goal (SDG) 7 of working toward clean and affordable energy by 2030 and for net-zero emissions by 2050. Palau is working on becoming energy independent, aiming to achieve its goal of 45% renewable energy by 2025 and 100% by 2032. Uchel recognized that achieving such progress would not be easy, especially with limited access to financing mechanisms, but confirmed that they are committed to pursuing the challenge.

Introductory remarks from Palau. Eden Uchel, director, Palau Energy and Water Administration, Ministry of Finance, stressed Palau's commitment to become energy independent through the use of renewable energy including ocean energy (photo by IMT-GT Joint Business Council).

Introductory Remarks by Romero Montenegro, Deputy Executive Director, Mindanao Development Authority, Philippines

Romero Montenegro highlighted that Mindanao's economy is driven mainly by agriculture and that marine resources, especially tuna fisheries and the culture of algae, have been a big part of that growth. He commented that global warming and climate change have already been felt in Mindanao with longer dry periods and a new normal of regular, stronger typhoons, but added that "there is no better time to act than now to pursue initiatives on energy transition and the blue economy." Montenegro detailed that new policies have been implemented in the past few years in the Philippines that support the transition to renewable energy and blue economy initiatives and revealed that he is working to establish a blue economy hub in Mindanao that will focus on marine pollution and climate change impacts, improving the sustainability of the ocean and water bodies and marine ecosystems by creating a pipeline of well-designed blue projects.

Introductory remarks from the Philippines. Romeo Montenegro, deputy executive director, Mindanao Development Authority, emphasized the urgency to pursue initiatives on energy transition and blue economy (photo by IMT-GT Joint Business Council).

Keynote Address
H.E. Rafizi Bin Ramli, Minister of Economy, Malaysia

Assalamualaikum wbt and a very Good Morning

Hon. Dato' Nor Azmie bin Diron
Secretary General, Ministry of Economy, Malaysia

Hon. Edi Prio Pambudi
Deputy Minister for International Cooperation, Coordinating Ministry for Economic Affairs, Republic of Indonesia

Romeo M. Montenegro
Assistant Secretary and Deputy Executive Director, Mindanao Development Authority (MinDA)

Alfredo Perdiguero, Director
Regional Cooperation and Operations Coordination Division, Southeast Asia Department,
Asian Development Bank

Firdaus Dahlan
Director, Centre for IMT-GT Subregional Cooperation (CIMT)

Eden Uchel
Director, Palau Energy and Water Administration

Distinguished Guests,

Ladies and Gentlemen,

At the outset, allow me to welcome all of you to the High-Level Investor Forum on The New Ocean–Energy Economy. I am pleased to see the participation of a number of thought leaders in the ocean–energy space, including business leaders, project developers as well as Provincial and Local Government authorities.

I would like to express my appreciation to the Asian Development Bank (ADB) and Centre for IMT-GT Subregional Cooperation (CIMT) for hosting such an auspicious event, which is also in conjunction with the 30th IMT- GT Anniversary Celebration.

This forum is organized to support and promote substantial cooperation for innovative and scalable integrated ocean–energy solutions and regenerative business activities in littoral states and communities.

With the theme "Introducing the New Ocean–Energy Economy," this platform can serve as an avenue for the global energy transition that will open up a spectrum of new opportunities for the long-term development of the country's economy, especially in new, value driven growth areas.

Distinguished Guests, Ladies and Gentlemen,

In Malaysia, the energy sector has long been a critical engine of growth for the economy. The energy sector will continue to play a key role in catalyzing socioeconomic development for the country and supporting the United Nations 2030 Agenda for Sustainable Development Goals (SDG 2030).

The National Energy Policy (NEP), 2022-2040, launched by Prime Minister of Malaysia on 19 September 2022 sets an overarching strategic policy direction for the country's energy sector, reaffirming our commitment for energy transition, to achieve the national goal of net-zero greenhouse gas (GHG) emissions by 2050 and to future-proof the energy sector.

The implementation of NEP will require collaboration and cooperation with a wide range of stakeholders in both public and private sectors, with new investments into emerging clean energy sectors.

Given the increased importance of environmental sustainability in the face of energy transition, amid many competing socioeconomic priorities, I believe there will be various low-carbon pathways. The journey, or the path, that each country takes depends on where their starting point is, what is their unique position, and what are the resources available and accessible to each.

Distinguished Guests, Ladies and Gentlemen,

The subregion of Indonesia-Malaysia-Thailand is uniquely situated in the Asia-Pacific region. At the centre of Southeast Asia, it is adjacent to the world's largest markets of the blue economy. Being a nation surrounded by seas, this subregion has tremendous growth opportunities to harness energy from the ocean. This is particularly important when there is great pressure to pursue a sustainable alternative energy.

Simultaneously, world energy consumption has been rapidly increasing and it is estimated to rise considerably over the next decades. Based on the World Energy Outlook published in 2012, electricity generating marine energy, including wave and tidal power, is expected to increase to 60 TWh (terawatt-hours) in 2035. Wave energy is enormous and more reliable than other renewable resources such as solar and wind energy because of its density (2-3 kW/m^2), which is greater than that of wind (0.4-0.6 kW/m^2) and solar (0.1-0.2 kW/m^2). Global wave energy resources are estimated at 17 TW h/year.

Distinguished Guests, Ladies and Gentlemen,

The power of ocean waves is truly amazing. Aside from thrilling surfing enthusiasts and enthralling beachgoers, their destructive potential has long earned the respect of generations of fishermen, and other mariners who encounter the forces of the sea. Yet, ocean waves can be harnessed into useful energy to reduce our dependency on fossil fuels. Instead of burning depleting fossil fuel reserves, we can obtain energy from a resource as clean, pollution free, and abundant as ocean waves. This is particularly important as Malaysia, during the 26th Conference of Parties (COP 26) in 2021 has committed toward being carbon neutral by 2050.

We are committed to developing renewable energy to reduce reliance on fossil fuels, resolve the increasing patterns of energy demands, and reduce greenhouse gas (GHG) emissions. In order to spur the growth of the renewable energy market, the government is encouraging renewable energy producers to apply for a Feed-in Tariff.

We are not insulated from global conditions. We experienced some of the most severe flooding in our history at the end of 2021 and the beginning of this year, which displaced about 100,000 people and caused a cumulative loss of 6.1 billion ringgit. Currently, higher input prices and disruptions in the food supply chain led to an increase in food inflation of 5.2% in May 2022. In the same way COVID-19 has had varying effects on different groups in society, and soaring food prices are hitting the B40 households even more than the rest of the population.

The challenges that we are facing now underscore the importance for all stakeholders in this subregion, including every level of government and the private sector, to adopt principles of inclusiveness and to pursue economic development sustainably. Addressing these interrelated challenges also requires innovative improvements and financing, technology sharing and capacity building. All of these endeavors are possible, but I would urge all participants to keep one central idea in mind today – that greater coordination of efforts will be required to support the transition toward a low-carbon economy. This is especially true within the relative complexity of the blue economy where, by their very nature, boundaries are more fluid, but it is encouraging that we are able to showcase so many exciting examples of such coordinated and integrated efforts today.

"The blue economy" is an important theme in the Twelfth Malaysia Plan that highlights our strategies to enhance the sustainability of our ports, shipping and marine transport, sectors which represent 40% of Malaysia's GDP. We also made a commitment to achieve net-zero GHG emissions by 2050 and to maintain at least 50% forest cover over the total land area.

To continue the central theme, linking the blue economy to Malaysia's Net Zero Commitments will require a coordinated regulatory and policy platform for maritime sector development.

Distinguished Guests, Ladies and Gentlemen,

The ocean is a vast and powerful resource that has the potential to revolutionize the way we generate electricity and drive economic growth. Wave energy, tidal energy offshore wind, ocean floating solar, and ocean thermal energy conversion are all promising forms of ocean energy that have the potential to produce large amounts of clean, renewable energy. We must continue to invest in research and development in these technologies to fully realize their potential and drive economic growth.

To this end, I believe that this subregion is well placed to emerge as a regional leader in low-carbon blue economy. I hope this event will serve as a platform for all participants to exchange ideas and share experiences that will contribute to the successful implementation of green growth policies and programs and promote climate resilience and prosperity in every sector and every part of our country. Let us work together to grasp the opportunities within our reach.

There is need of high-level investors to drive the development of ocean energy in these subregional areas that includes:

- bridge the gap between research and commercialization;
- to build the necessary infrastructure; and
- to create the jobs and economic opportunities that ocean energy can bring.

Therefore, I hope that this forum can reiterate the importance of establishing strong partnerships in this region and highlight the mechanisms through which stakeholders may access support. We all have a role to play in the new ocean energy.

There are many short-term challenges ahead of us. However, so long as we steer the course and remain committed to the energy transition process, we will reap its benefits in due time.

I look at you, the energy experts to take on some of these challenges. I forward to what you have to say.

Distinguished Guests, Ladies and Gentlemen,

On that note, I wish you a successful and productive forum.

Thank you.

Ministry of Economy
Putrajaya

7 February 2023

Plenary 1: The New Ocean–Energy Economy

Opening Remarks

Netty Muharni, Assistant Deputy Minister for Regional and Sub Regional Economic Cooperation, Coordinating Ministry for Economic Affairs of Indonesia (recorded video)

Netty Muharni observed that for countries with coastal areas, offshore renewables can create new jobs and reduce carbon dioxide production, both of which are goals of Indonesia, the ASEAN, and the UN SDGs. She acknowledged that at this time, Indonesia's alternative energy production is quite small, only 3%. As the world's largest archipelago, though, Muharni was keen to focus on Indonesia's huge potential for alternative ocean energy, which is around 216 GW. She considered the significant barriers to making progress include higher cost of investment; lack of good data, for example on wave and tidal current energy in many potential areas; and lack of legislation to support alternative energy. More positively, she reported that Indonesia has already started to plan for alternative ocean–energy production in several areas using both waves and tidal currents, and said that the government can create new policy, and consumer preference can create a new market. She closed by saying that "Alternative ocean energy will support the goals of net-zero sustainable development and provide new jobs and a longer supply chain, but the development requires high-cost investment to achieve inclusiveness and sustainable development."

"Under our subregional economic cooperation, ocean energy could play a role in supporting the goals of IMT-GT sustainable urban development framework by providing a clean and renewable source of energy for the regional growing population and economies"

Plenary 1 opening remarks. Netty Muharni, assistant deputy minister for regional and sub regional economic cooperation, Coordinating Ministry for Economic Affairs of Indonesia shared that Indonesia has already started looking at ocean energy production to harness the country's vast ocean energy resources (photo by IMT-GT Joint Business Council).

Introduction to the New Ocean–Energy Economy

Nick Lambert, Founder, NLA International Ltd

Nick Lambert, a retired rear admiral, explained the various sectors involved in MARES. As the acronym indicates MARES includes alternative energy such as OTEC and green hydrogen, marine aquaculture, coral reef rehabilitation, coastal protection, artificial reefs, and marine ecotourism; and some of the important connections between them. The presentation emphasized the following:

- The answer to climate change and to social and economic benefits lies in the oceans. A profit must be made, and cash has got to flow. But historically, things have been taken out of the ocean, made stuff out of it, and then throw the waste back in the sea. The world is still too tied to a linear, extractive and destructive approach and this needs to be changed. Anything that is done at sea should be undertaken in a regenerative way—i.e., whatever is taken to get social and economic benefits from the sea space has to be done in a regenerative way, meaning it is in a better state than when it started.

- Countries, like Maldives, should be referred to as "large ocean states" and not as "small island developing states." These large ocean states are the flag bearers because they are right on the front line of the fight against climate change. They did not cause it, but they are showing the opportunities to make a difference.

- There is only one ocean—not five oceans and seven seas—because they are all interlinked. The ocean is like a machine, and it works, and it keeps us alive, allowing us to breathe and to eat.

- Marine renewable energy is at the heart of the opportunity to reverse climate change. If offshore wind production were to reach its peak potential, it could generate more than 120,000 GW, which is 11 times the global energy demand in 2040. However, there is a need to move faster. The International Energy Agency reported that in order to achieve the net-zero target by 2050, the use of marine renewable energy should be increased by 33% per annum. Currently, marine energy contributes only to 13%. All the marine power is out there, but nothing is being done. It is not being exploited at it should.

- One factor that stops the development of marine renewable energy is that many island nations that could benefit from renewable energy do not actually require a lot of it for their basic needs. As such, there is a need to be innovative and think about multifunction and look at the hydrogen opportunity.

- The technologies are already here and ready to be deployed. Almost all of them are at a high-technology readiness level. Particularly considering the increase in cost of energy over the last year, they are now viable putting out kilowatt-hours at an affordable price. The issue is simply how to scale them up.

- The Shetland Islands in the United Kingdom are a great example of how this transition from a carbon-based economy can be shifted to a hydrogen-based economy using Integrated Marine Spatial Planning. They are a tiny Council, with only 29,000 people in their population, but their vision is to use wind power to generate electricity and to make hydrogen, which will be used to run boats and other equipment, with the excess power and hydrogen being sold for a profit.

- We are now getting to the point where marine renewable energy is potentially a very good investment, whether it is solar or OTEC. If we can invest and implement these technologies in a circular, regenerative, and long-term manner, then we can make them work to reverse climate change. But we must be ambitious and drive toward multifunction marine operations at scale.

"During my early days as an ordinary seaman, we used to throw all our trash over the side of the boat without a thought to where it might end up. We've moved on so far from that point, but the world is still too tied to a linear, extractive and—dare I say—destructive approach. We need to change that so that anything that we do at sea should be undertaken in a regenerative way: whatever I do to get social and economic benefit from this sea space, I will do it in such a way that I will leave it in a better state than when I started."

Introduction to the New Ocean–Energy Economy. Nick Lambert set the stage for more discussions about the topic (photo by IMT-GT Joint Business Council).

Panel Discussion: Integrating the New Ocean–Energy Economy to Global Energy Needs

Panel Discussion: Integrating the New Ocean–Energy Economy to Global Energy needs. Cindy Cisneros-Tiangco, principal energy specialist, ADB facilitated the panel discussion (photo by IMT-GT Joint Business Council).

Marine renewable energy and its economics were discussed by the first panel of the forum titled Integrating the New Ocean–Energy Economy to Global Energy Needs, moderated by Cindy Cisneros-Tiangco, Principal Energy Specialist, Pacific Department, ADB with

- James Doldolia, Program Coordinator, Mindanao Development Authority (MinDA) Power Development Program
- Kee-Yung Nam, Principal Energy Economist, SDSC-ENE, ADB
- Michael Abundo, CEO, OceanPixel Pte Ltd
- Wan Sayuti Wan Hussin, Head (Strategy and Policy), Corporate Sustainability, Petronas Sustainability Unit, Petronas BHD

Cindy Cisneros-Tiangco said that ADB has the ambition to become the region's climate bank with a target of $100 billion in climate change adaptation and mitigation through 2030, and that they are looking for sector integrated solutions.

Wan Sayuti reported that, in 2020, oil and gas company Petronas announced that its goal was to become net zero in carbon by 2050. So far, its corporate social responsibility projects in Malaysia include mangrove and seagrass replanting, and rigs-to-reefs for decommissioned oil rigs.

James Doldolia highlighted that the Philippines is one of the top three producers of algae globally. Most of this is farmed in the southern parts of the country, especially in Tawi-Tawi. He pointed out that seaweeds can sequester a lot of carbon, but global warming is affecting the commercial production of seaweeds and so there is a lack of freshwater for processing them. Doldolia also revealed that the Mindanao Development Authority has launched a new project involving hybrid solar power plant so that the power can be used to run a desalination plant to produce fresh water and for other uses.

Michael Abundo queried where the gap is between existing technology and putting this technology to work. He considered the first gap to be an "awareness" gap. Many people are still not aware of the many types of renewable ocean–energy technology. Other gaps he highlighted were capacity building and capability of implementation and finding the right level of funding for each project. He concluded by saying, "That is why this ADB MARES HLIF is so important, because we have the right mix of people in this room to make things happen."

Kee-Yung Nam stated that the ADB's energy sector program is assessing green hydrogen, floating solar, and offshore wind—"the benefit side of the blue economy"—and linkages to the market economy. He reflected that the economic efficiency of each technology also needs to be tested and that for a successful energy transition to be realized, it would need to be clearly identified which type of clean energy is the most efficient and least-cost option. He also noted that where market failures occur, ADB can step in "to maximize benefit given limited resources."

Cindy Cisneros-Tiangco summed up by commenting that ADB is "very risk averse" and identifying that there are still a lot of gaps within the governments, who are creating the regulatory framework and requesting implementation of the various renewable ocean–energy technologies in their countries. These challenges were worth addressing though. In closing, she said, "The oceans will save the planet and us."

Presentation of Eight Shortlisted MARES Projects

Andy Hamflett, Co-Founder and Director, NLA International Ltd. presented eight short videos that highlighted the key elements and benefits of the projects that had been shortlisted as part of the MARES global competition to find innovative multifunction projects. These were the following:

Blue-Cooking

The Blue-Cooking model looks to utilize electricity generated from floating solar PV panels integrated into aquaculture farms in order to enable the introduction of clean cooking (using electric cookstoves rather than solid fuels). Initially focusing on three communities in Bangladesh, the project will work with aquaculture farmers to understand the barriers and benefits of floating solar PV system integration, including mechanisms to improve income and farming yield. The project will also work with women to understand socioeconomic barriers and benefits to the adoption of clean cooking, in addition to transitioning to new technologies and wider health improvements.

Decom2Green

The Decom2Green project aspires to catalyze the regenerative decommissioning of offshore platforms within Asia. It aims to engage with oil and gas operators in Thailand to explore the feasibility of repurposing decommissioned offshore platforms to allow them to support aquaculture and marine renewable energy operations. This innovative approach promises benefits both to the energy companies (by reducing the costs of decommissioning and strengthening company's environmental, social, and governance credentials) and coastal and islands states (by enabling new, regenerative blue economy industries that strengthen food and energy security and providing jobs).

Deep Water Intake Infrastructure Powered By Megawatt-Scale OTEC

An industrial academic consortium from Japan is looking to harness the potential of deep open water and OTEC to enable communities to create their own local power, farm, and fish more reliably, utilize cool buildings with less electricity, and provide environmentally friendly desalination for water security. The technological approach, which has been deployed successfully in other parts of the world, seeks to set up demonstrations for the first time in Palau, and will use the cold, clean, and nutrient salts from deep water to provide a range of regenerative benefits to local communities.

Ongedaol Nature Resort Palau

Ongedaol Nature Resort Palau seeks to create an innovative new model for sustainable, organic ecotourism, empowered by modern technology, that is completely off-grid and formed in partnership with local landowners. A total of 23 eco-retreat suites amid a natural reserve will feature sustainable organic farming (agriculture and aquaculture) where visitors can experience living amid the jungles of Palau, taste locally grown food, hike ancient trails, and enjoy nature on both lands and in the water. From ownership structure to design, the project will support local livelihoods, utilize renewable energy systems, and create a unique eco-friendly experience inspired by Palauan traditions.

Pacific Ocean Explorers

The Pacific Ocean Explorers project would combine marine renewable energy and aquaculture to strengthen food and energy security in Palau. By establishing the natural spawning cycles of certain species of significance, eggs can be harvested to produce juveniles for food grow out, restocking for conservation, and restocking for food security. All of this would be achieved while using 80%–100% renewable energy provided by OTEC, as well as providing staff and other resources to support coral propagation for reef resilience and to combat the effects of human-induced impact.

Republic of the Marshall Islands Climate Proof Fuel Storage

This project aims to use converted maritime tankers for floating fuel storage, which will replace a land-based system that is vulnerable to sea level rise and more severe meteorological events exacerbated by climate change. Used tankers will be retrofitted by the Marshalls Energy Company in accordance with industry best practices. They may also be embellished by renewable energy (e.g., solar, wind, wave, and/or tidal energy conversion systems) to support the charging of electric and hybrid boats and to produce hydrogen, ammonia, and other high-value chemicals, which can be sold into existing markets.

Savusavu Blue Town Model

The Savusavu Blue Town Model looks to introduce a comprehensive framework to create thriving communities that address climate change and ocean pollution while building sustainable livelihoods within a picturesque harbor town in Fiji. Savusavu provides a safe anchorage for visiting yachts and is the focus of tourism activities on Fiji's second largest island of Vanua Levu. By adopting a sustainable and community-driven approach to planning, the Savusavu Blue Town Model aims to develop a master plan that will strengthen the local economy at the same time as enhancing marine conservation, supporting environmental education, and transitioning to marine renewable energy.

Subic Blue

The Subic Blue project aspires to introduce the first aquarium fish farm and coral nursery in the Philippines, producing large quantities of tropical fish for sale and for placement back onto protected reefs. In addition to bringing a unique regenerative business to the education-focused Subic Bay Marine Exploratorium, Inc. (SBMEI) on the west coast of the island of Luzon, the new operations would also create additional jobs and provide solar energy for SBMEI and the local grid. Planned activities would also rehabilitate local coral reefs and provide real-life marine education to thousands of tourists and students from the Philippines and overseas.

Plenary 2: Using Energy from the Ocean

Opening Remarks

Thuttai Keeratipongpaiboon from the Office of the National Economic and Social Development Council (NESDC), Kingdom of Thailand, delivered the opening remarks which set the scene for the inclusion of all sea-space users. He explained that Thailand has integrated renewable energy into its National Energy Plan coming into force this year and this includes a goal to achieve decarbonization, digitalization, decentralization, and electrification. Keeratipongpaiboon also commented that Thailand has committed to a program of transitioning 30% of all vehicles to electric vehicles by 2030 and that Thailand is committed to a blue economy, even it is still in the R&D phase.

"Thailand has established the Southern Economic Corridor in the strategic southern coastal provinces, specifically targeting tourism while conserving marine resources. This will enable us to promote economic growth while simultaneously safeguarding and restoring the ocean's health."

Opening Remarks, Plenary 2. Thuttai Keeratipongpaiboon, Office of the National Economic and Social Development Council, Thailand mentioned that the country's ocean energy is still in the R&D stage and that the involvement and cooperation of stakeholders in addressing this matter further is welcome (photo by IMT-GT Joint Business Council).

MARES Project Competition Results

Jonathan Turner, Director, NLA International Ltd. revealed that the winning project – which best reflected the overall multisector goals of MARES – was Subic Blue. It was selected as it demonstrated strong MARES thinking on integrated multifunction development, with marine renewable energy as a strong element. It was inclusive, with a strong gender component. Judges had considered that the project displayed impressive capability development and education components, with the potential to train people up in other locations. Finally, it convinced judges that it was scalable, with a strong market case delivered.

Turner related that the judges found the decision-making very difficult, as each of the eight finalists had their own merits. However, the judges asked that honorable mentions were given to three additional projects that demonstrated the multifunction approach and also showed clear benefits to all sea-space users. These were Pacific Ocean Explorers; Deep Water Intake Infrastructure powered by MW-scale OTEC; and Blue-Cooking. The brief descriptions of the eight finalists are included in Annex 1.

Presentations

Integration of the Hydrogen Economy and Marine Renewable Energy

Ines Marques, Director, Green Hydrogen Development Plan, Green Hydrogen Organisation, stated that we have the technology, we have the potential, but we do not have the scale. What we need is public policy to support implementation and public-private finance for projects, and to make sure that local communities benefit from green hydrogen. She said that she really appreciated the MARES approach to think about this holistically, to think about how we can use offshore wind floating solar OTEC to produce green hydrogen and its derivatives, but also to think about how it can link to other industries. She suggested including production of green hydrogen, green ammonia, and green fertilizers for aquaculture and agriculture; or producing green hydrogen and green fuels for the maritime industry and decarbonized shipping.

Dolf Gielen, Senior Energy Economist, World Bank clarified that about 100 million tonnes (Mt) of hydrogen is produced and consumed per year globally. He said that hydrogen is a mature industry, but that only about 2% is clean hydrogen: 1% green, 1% blue, and 98% grey hydrogen from natural gas and coal. Gielen stated that only Gigawatt electrolyzers are in operation; the price of green hydrogen is $4–$5/kg and costs should drop rapidly. He stressed the need to develop the production, transport, and demand. He also highlighted that there are >500 hydrogen projects planned worth $240 billion but that very few of these are in developing countries.

Petra Schwager, Global Program for Green Hydrogen in Industry, Climate and Technology Partnerships Division, UNIDO, said that green hydrogen projects are large, starting at $50 million and up. She also commented that UNIDO is working to de-risk investments in green hydrogen production, create standards, and ensure that projects also include social incentives for local communities, such as more access to clean fresh water.

Jonas Moberg, Green Hydrogen Organization, declared that the mission of the Green Hydrogen Organization (GH2.org) is to dramatically accelerate the production and utilization of green hydrogen across a range of sectors globally. The company intends to push to rapidly decarbonize industries like steel, cement, fertilizers, shipping, and aviation, the sectors that have so far made limited progress reducing their emissions. He also emphasized the need for governments to facilitate green hydrogen projects through policy and regulation.

Panel Discussion: Integrating Micro, Small and Medium-Sized Enterprises into the New Ocean–Energy Economy

The second panel, titled "Integrating Micro-, Small-, and Medium-sized Enterprises (MSMEs) into the New Ocean–Energy Economy," discussed transition, engaging with stakeholders, and how larger infrastructure developments can make space for smaller sea-space users.

Panel Discussion: Integrating Micro, Small, and Medium-sized Enterprises into the New Ocean–Energy Economy. James Elsmoor, founder and CEO, Island Innovation, LLC moderated the second session (photo by IMT-GT Joint Business Council).

The moderator, **James Ellsmoor, Founder and CEO, Island Innovation LLC,** reflected that the forum had heard a lot about islands during the proceedings, and that he especially liked the name large ocean states instead of small island developing states. He then asked each of the panelists to address this question: "What are the big opportunities for MSMEs in island economies?"

Dan Millison, Consultant, SDSC-ENE, ADB, thought that there is definitely a big role for small- and medium-sized enterprises (SMEs) to play in the New Ocean–Energy Economy, for example through the use of global energy for 'power-to-x' value addition. An offshore wind developer probably has no interest in marine aquaculture, he suggested, but many things will be growing on whatever we place in the sea—for example, a pontoon for an offshore wind structure. He reported that there are already small recreational fishing boats visiting offshore wind farm facilities because the fishing is better there because of all the additional growth of marine life on the structures. Millison suggested that governments can require wind farm developers to include aquaculture development by SMEs in their wind farm developments, stating that this type of growth occurs everywhere in the world and aquaculture is also possible in any marine area used for wind farms. He also highlighted another way of looking at finance, stating that, while the main business of development banks is in making large sovereign loans to countries, ADB can loan to intermediary banks, which in turn can loan to SMEs.

Irene Olkeriil, Founder, EmeralDreams Services, LLC, shared that she is an entrepreneur at heart and posed that for any technology to be successful, it is important to include the people. She said that in Palau, we are people of the sea, and we welcome Blue Ocean technologies. Olkeriil said that the people and the culture are very important and encouraged the audience to remember that 40% of tourism is cultural.

Salma Khoo, from nonprofit Jaringan Ekologi Dan Iklim, told the panel that she is an islander from Penang, representing a nonprofit SME working with coastal communities. She pointed out the importance of integrating local people's activities, especially fishing, into the planning process for new Blue Ocean energy technologies. For example, there are lots of different types of fishermen, and they use different fishing methods in different parts of the coastal and marine space such as intertidal mudflats, mangrove forests, seagrass beds, and coral reefs. Therefore, it will be essential to consider each of these activities when designing the installation of new technologies such as hydrogen infrastructure. Khoo mentioned FAO's good guidelines on small-scale fisheries. Small-scale fishing or inshore fishing are very sustainable and circular economy. Her organization wants to help inshore fishermen, especially their ecosystems. The sea is the shared commons for the coastal communities, who are called as the stewards for the biodiversity of the sea. With respect to integrating new technology so that it can benefit the local communities, she recommended getting the fishermen's cooperatives involved in aquaculture. This would be useful because they understand the monsoons and tidal fluctuations. Reflecting that they have seen environmental impact assessment reports that assume static conditions in the sea when in fact, the sea state is constantly changing. Consulting local fishermen could help to reduce risk.

Sarah Dole, Founder, Invena, thought that there are vast opportunities to tap into in the new Blue Ocean Economy, to help reduce the risk in the face of rapidly rising sea level and changing shorelines in island states. She observed that shoreline protection structures often fail in the face of severe ocean conditions such as storm surge and large waves, threatening local communities. As new structure protection, such as sea walls, and new land by reclamation can often lead to failure. She highlighted one of the approaches they are taking to build low-profile structures in shallow waters of the reef platform that naturally accumulate and build up sand berms behind them just by the flow of water over them. She further explained that they then plant mangroves to stabilize these new structures. Once stabilized and colonized, the new structures form part of the natural coastal protection system, protecting coastal communities, and form a new habitat for marine organisms that may be of interest for tourism and/or fisheries. Dole said that, as a physicist by training, she was pleased that artificial intelligence is being used to analyze the bathymetric and water flow data to help design and place new sand accumulation structures in the Maldives lagoons. This technology should be useful in many other coastal areas around the world, she concluded.

Scott Countryman, Executive Director, The Coral Triangle Conservancy, disclosed that he had started out as an entrepreneur in business and technology. After 25 years in the Philippines, he realized that business success was not enough, and that it is important to have environmental considerations because many of the beautiful natural reefs he had been visiting were dying. He applied the ideas from business and used them to motivate local people to conserve their ecosystems. This led him start the Coral Triangle Conservancy to setup networks of marine protected areas. He was able to apply the same technologies that he was using for his online businesses and digital media for dynamite detection, fishpond, detection, etc. They started to plant corals one by one to rehabilitate coral reefs, but they discovered that the larger problem relate to fish farming, unsustainable fishing methods, or overfishing from commercial vessels. They also noted that the biggest challenge was how to value the coral reefs. Thus, they developed a methodology using artificial intelligence, block chain technology, etc. to put a value on the natural assets as a way of natural assets accounting. Through their organization, both poverty and malnutrition issues are being resolved by providing jobs for people to restore the ecosystems. They are successful in bringing together communities that are now financially motivated to improve their quality of life through off-grid energy, as well as access to education, communication, and electric power boats.

Plenary 3: Responsibly Using Marine Spaces

Andreas Dipi Patria, Coordinating Ministry of Maritime and Investment Affairs, Republic of Indonesia, provided opening remarks, which highlighted his country's action on maritime investment. With more than 17,500 islands and 108,000 kilometers of coastlines, and three-quarters of its territory at sea, oceans are central to Indonesia's prosperity. Yet there are challenges to the extent and integrity of Indonesia's marine and coastal ecosystems, which threaten its economic value, starting from overfishing, degradation of mangroves and coral reefs, and marine debris. These challenges can be addressed through a blue economy, or a sustainable ocean economy strategy. A blue economy creates healthy oceans, resilient coastal livelihood, and sustainable economic growth. The blue economy creates a healthy ocean and stable economic growth. Indonesia's 2045 Blue Economy Vision is the basis for utilizing marine resources. Indonesia has acknowledged the value of the blue economy as its main reference in achieving a sustainable ocean for the greatest prosperity of the people.

Panel Discussion: Financing the New Ocean Energy and Blue Economies

The moderator, Ghislain De Valon, Lead, Blue Southeast Asia Finance Hub, ADB, opened the discussion by asking what the biggest challenges are for financing MARES-type projects

Len George, Principal Energy Specialist, Energy Division, Pacific Department (PAEN), ADB commented that ADB has about 14 countries in the Pacific Department and that, of the 16 SIDS or large ocean states, 14 are in the Pacific region. These countries have high electricity costs, and most of their power is generated using diesel fuel. Over the past 3 years, fuel prices in these countries have more than doubled. Governments in these countries have created very high targets for decarbonization; but, if working with small grids, the ability to connect them with renewable energy sources can be challenging from an engineering point of view. Private companies often approach the energy providers with renewable energy "deals" to provide energy at lower costs, but this is challenging as these proposals should have good technology that can match with the grid, and that the proposed pricing is realistic. Many of these countries have very large marine areas and, while there are some examples of green hydrogen being exported out of developed countries such as Chile and Australia, it is still a question whether small island countries with limited technical capabilities can replicate this. George enumerated a range of challenges that need to be addressed: setting the vision properly, focusing on sustainable supply, defining what public sector support will be needed, and questioning if the competitive advantage is "for real." The biggest challenge in finding financing lies in determining how well the technology fits—sometimes decision paralysis sets in if there are too many options.

Joan Fulton, COO, Ocean Assets, introduced her company by pointing out that Ocean Assets' mission is to scale impact, using private capital for Blue SMEs in Asia and the Pacific through joint collaboration with ADB, the United Nations Development Programme, and UNEP. With a focus on matching investors with SMEs, she said, they must ask: "What do investors want?" Fulton listed five main characteristics that investors focus on: geography, level of economic development, impact vs. return trade off, size (asset heavy or light), and which blue economy sector. She went on to list several additional considerations which included old vs. new technology; large size, SME, or micro; technology readiness level (TRL), and finally, what is the growth potential—high vs. low. These considerations may also differ, depending on the type of investors involved: impact funds, wealth managers, family companies, foundations, angels, friends, insurance companies, development finance institutions, or pension funds. Fulton considered that public–private partnerships are essential for catalytic financing and that investors are interested in investments where the risks have been reduced by various means such as insurance.

Panel Discussion: Financing the New Ocean Energy and Blue Economies. Ghislain De Valon, lead, Blue Southeast Asia Finance Hub, ADB led the discussion about the challenges in financing MARES-type projects (photo by IMT-GT Joint Business Council).

Michael Dethlefsen, Chief, Innovative Finance and International Financial Institutions Division, UNIDO, explained that UNIDO is a small UN organization based in Vienna with operations around the globe working on SDG 9 (Industry, Innovation and Infrastructure) and overlapping with SDG 14 (Life Below Water), to promote sustainable industrial development in 171 member states. UNIDO is trying to go beyond technical assistance, agreements, and partnerships to promote actual investments in SMEs. For example, UNIDO has been trying to help develop and implement green hydrogen projects in developing countries and is trying to work more with local banks to invest in SMEs. The biggest challenge for renewable ocean energy implementation is in providing assistance to SMEs and the skills sets for learning new qualifications that are needed to make it happen. UNIDO uses a cluster approach to bring together community members to work together to identify the best way forward to get started and scale up. UNIDO wanted to do more to match investors with SMEs.

Alix Burrell, Principal Investment Specialist, Private Sector Operations Department, ADB, shared that there is a lot of money available for financing, ocean technologies exist, and they are scalable. But there is a gap in matching the projects with financing. She explained that equity investors are optimists because they believe a project will be profitable, whereas impact investors are the pessimists who believe in creating social and environmental impacts and hope to get their funds back eventually. So, the lesson here is to get the equity [startup investments] in place as soon as possible, as lenders do not lend to people without any money or equity. She highlighted that a business plan must be developed first and thought that ADB might be able to help mobilize other private sources of funding for the project. Key questions to address were: What is your revenue stream? How and when will your investors start to make money? How will you convince investors to scale up the business?

Burrell underlined that aside from assisting with mobilizing funding from other sources, ADB has a private sector division, which can provide funding. Crucially, though, the amount of work and time needed to fund a $5 million project is about the same as what is needed for a $50 million project, and the proponents of the $50 million project are likely an existing large company with a track record. Therefore, larger projects will be more attractive to ADB and many impact investors. She commented that ADB therefore needs to do more to help fund smaller $5 million MARES-type multisector projects and thought that there are ways to do this within the system. Burrell suggested that they could use the revenue stream from alternative energy projects to fund co-located projects such as aquaculture and/or ecotourism, and that they could look for role models for scaling and to achieve impact and save the planet. She ended with some advice for project owners: lenders look at risks, and these projects are really exciting, but do not use the same pitch with all types of investors. She added that a good business plan is essential.

Vincent Choy, Asia Representative, Delphos, explained that Delphos is doing a lot of fundraising in the region and that they are facilitating SMEs to get funded in the range of $30 million to $400 million. He revealed that he himself had started as an entrepreneur and has been fundraising throughout his career. Reflecting that there are so many different funders around the world, he commented that it is important to remember that each funder will be interested in different aspects of your project. Thus, it is usually necessary to create several pitch decks that match what each particular funder may be interested in. He shared that for one his own companies, he might have as many as 18 different pitch decks carefully designed for 18 different potential funders' interests. If there is an opportunity to make a pitch, it is essential to ask the funder some questions to find out what they are looking for to pull out the best deck to match the potential funder's interests. Choy stated that the first two minutes of any pitch are critical and suggested to be imaginative in figuring out various ways to make money from the same project.

Summary of MARES Progress and the High-Level Investor Forum

Remarks by Steve Peters, ADB

Steve Peters, Energy Specialist at ADB, highlighted the following:

- The key message is that we need to love the ocean more by putting back what we have taken and think ways how to make money out of it.
- The MARES technical assistance (TA) project and the eight projects that were presented attested that renewable energy, specifically the marine renewable energy, has already made significant advancements.
- ADB, through the MARES TA, has undertaken extensive research on marine renewable energy and developed business planning models for the Marshall Islands and Palau for possible financing.
- The ADB MARES data room has a list of webinars and knowledge products developed through this TA. One major output of the TA is the handbook that provides an overview of how coastal and island states can adopt a multifunction approach to protect and regenerate the marine estates and exclusive economic zones (EEZs), while deriving social and economic benefits.

Summary. Steve Peters, senior energy specialist, ADB provided the key takeaways of the Forum (photo by IMT-GT Joint Business Council).

We can love the ocean more and put more into the ocean than we take out and we can make money. We've done 2 years of research. We have two 400-page business planning models for the Marshall Islands and Palau. We need to look at business models and pathways for financing. If we want to beat climate change, we need to do it all together."

Closing Remarks and Call to Action

Comments by Priyantha Wijayatunga, Chief of Energy Sector Group, SDCC, ADB

Priyantha Wijayatunga, noted the following key words that came out of the forum, which include extract, use, dispose; not only sustainable but regenerative; circular; multifunction; not a boundary but an interface; just ocean; small island states are large ocean states; action is attraction; backward and forward linkages in the blue economy; good business plans; combining sovereign with nonsovereign; the future is already here, it is just not evenly distributed.

He stressed that the event touched on all five alternative energy principles: providing clean, reliable, affordable energy to people in ADB DMCs; well invested; resilient; good governance; knowledge/cross-sector regional integration.

In closing, Wijayatunga announced that ADB is currently reorganizing to focus on solutions not just projects. All sectors will come under one umbrella where strong interaction with private sector and public–private partnerships are envisaged. He assured that ADB's Energy Team will make the best efforts to make it happen.

Closing remarks. Priyantha Wijayatunga, chief of the Energy Sector Group ADB, provided the concluding remarks (photo by IMT-GT Joint Business Council).

Annex 1: Project Competition Proposals

The four-page competition proposals for each of the projects can be found at the links below.

The winning project:

Subic Blue. The Subic Blue project aspires to introduce the first aquarium fish farm and coral nursery in the Philippines, producing large quantities of tropical fish for sale and for placement back onto protected reefs.

Judges' honorable mentions

Pacific Ocean Explorers. The Pacific Ocean Explorers (POE) project would combine marine renewable energy and aquaculture to strengthen food and energy security in Palau.

Blue-Cooking. The Blue-Cooking model looks to utilize electricity generated from floating solar energy panels integrated into aquaculture farms in order to enable the spread of clean cooking (using electric cookstoves rather than solid fuels).

Other shortlisted projects:

Decom2Green. The Decom2Green project aspires to catalyze the regenerative decommissioning of offshore platforms within Asia.

Deep Water Intake Infrastructure powered by MW-scale OTEC. This project would harness the potential of Deep Open Water (DOW) and Ocean Thermal Energy Conversion (OTEC) to enable communities to create their own local power, farm, and fish more reliably, cool buildings with less electricity, and provide environmentally friendly desalination for water security.

Republic of the Marshall Islands Climate Proof Fuel Storage. This project aims to use converted maritime tankers for floating fuel storage, which will replace a land-based system that is vulnerable to sea level rise and more severe meteorological events exacerbated by climate change.

Ongedaol Nature Resort Palau. Ongedaol Nature Resort Palau seeks to create an innovative new model for sustainable, organic ecotourism, empowered by modern technology, that is completely off-grid and formed in partnership with local landowners.

Savusavu Blue Town Model. The Savusavu Blue Town Model looks to introduce a comprehensive framework to create thriving communities who address climate change and ocean pollution while building sustainable livelihoods within a picturesque harbor town in Fiji.

Annex 2: Forum Organizers

Asian Development Bank

ADB is committed to achieving a prosperous, inclusive, resilient, and sustainable Asia and the Pacific, while sustaining its efforts to eradicate extreme poverty. Established in 1966, it is owned by 68 members —49 from the region. Its main instruments for helping its developing member countries are policy dialogue, loans, equity investments, guarantees, grants, and technical assistance.

Centre for Indonesia–Malaysia–Thailand Growth Triangle

The Centre for Indonesia–Malaysia–Thailand Growth Triangle (CIMT-GT) is the secretariat for Indonesia–Malaysia–Thailand Growth Triangle (IMT-GT) and the Asian Development Bank. The IMT-GT subregional program aims to stimulate economic development in 32 of these three countries' less-developed states and provinces. ADB has been involved in the IMT-GT program since its inception and has been a regional development partner since 2007.

Indonesia–Malaysia–Thailand Growth Triangle Joint Business Council

The IMT-GT Joint Business Council serves as the focal point of the private sector with a clear mission of encouraging the private sector to pursue trade and investment opportunities. It aims to enlarge its membership base to include micro-, small-, and medium-sized enterprises; small- and medium-sized enterprises; social enterprises; and high-caliber players and representatives of large corporations.

NLA International Ltd.

NLA International Ltd. champions the implementation of blue economy solutions through the utilization of innovative technologies, tools, and processes to create sustainable ocean environments for the people and economies that depend upon them. The NLA International Ltd. team comprises a diverse skill set with a breadth of experience from in and beyond the blue economy.

Underpinning Assumptions for Indicative MRE Estimates for the Marshall Islands

These estimates relate to Table 5 on p. 69.

(a) The area of computation for the low scenario is 1% of the total available area (10% of the total available area for the High Scenario). For tidal/current, low scenario is 0.5 m/s in 0.1% of the total available area, while high scenario covers velocities 1 m/s in 1% of the total available area.

(b) The perimeter of the EEZ of Marshall Islands is 1,603,536.43 km^2 (EEZ US and associated territories data from http://geo.pacioos.hawaii.edu/geoexplorer/). The value for the EEZ no-take zone is 80% of the EEZ.

(c) The rated output used are the following: marine solar energy = 7 m^2/kW; wave energy = 10 MW/km; OTEC = 0.8 MW/km^2 (https://library.oapen.org/bitstream/handle/20.500.12657/43849/external_content.pdf?sequence=1); offshore wind = 310.69 W/m^2 (wind atlas @ 100 m; globalwindatlas.info); marine biomass = 2,500 MWh/km^2-yr (annual yield); tidal or current = 42 MW/km^2.

(d) The coastline length is 370.4 km (https://www.cia.gov/the-world-factbook/field/coastline/). The final value used is the value for the low scenario.

(e) The final value used is the value for the low scenario.

(f) International Renewable Energy Agency. 2022. *Global Hydrogen Trade to Meet the 1.5°C Climate Goal: Part III – Green Hydrogen Cost and Potential.* Abu Dhabi.

(g) S. Banerjee, M. N. Musa, and A. B. Jaafar. 2017. Economic Assessment and Prospect of Hydrogen Generated by OTEC as Future Fuel. *International Journal of Hydrogen Energy.* 42 (1). pp. 26–37. https://doi.org/10.1016/j.ijhydene.2016.11.115. *(Note there is limit references to 2030 H$_2$/kg pricing in literature. Pricing for marine renewables and hence H$_2$ have seen a significant reduction.)*

(h) World Bank. *2021. Offshore Wind Roadmap for the Philippines.* Washington, DC.

(i) N. Dinh. 2022. *Projections of Levelized Costs of Hydrogen (LCOH), MARES Report.* Manila: ADB.

(j) Salinity gradient research and technologies in the Marshall Islands need more exploration, investigation, and feasibility studies.

Underpinning Assumptions for Indicative MRE Estimates for Palau

These estimates relate to Table 6 on pp. 78–79.

(a)   The area of computation for the low scenario is 1% of the total available area (10% of the total available area for the High Scenario). For tidal/current, low scenario is 0.5 m/s in 0.1% of the total available area, while high scenario covers velocities 1 m/s in 1% of the total available area.

(b)   The perimeter of the EEZ of Palau is 475,077 km^2 (EEZ US and associated territories data from http://geo.pacioos.hawaii.edu/geoexplorer/). The value for the EEZ no-take zone is 80% of the EEZ.

(c)   The rated output used are the following: marine solar energy = 7 m^2/kW; wave energy = 10 MW/km; OTEC = 0.8 MW/km^2 (https://library.oapen.org/bitstream/handle/20.500.12657/43849/external_content.pdf?sequence=1); offshore wind = 114.57 W/m^2 (wind atlas @ 100 m; globalwindatlas.info); marine biomass = 2,500 MWh/km^2-yr (annual yield); tidal or current = 42 MW/km^2.

(d)   The coastline length is 1,519 km (https://www.cia.gov/the-world-factbook/field/coastline/). The final value used is the value for the low scenario.

(e)   International Renewable Energy Agency. 2022. *Global Hydrogen Trade to Meet the 1.5°C Climate Goal: Part III – Green Hydrogen Cost and Potential*. Abu Dhabi.

(f)   S. Banerjee, M. N. Musa, and A. B. Jaafar. 2017. Economic Assessment and Prospect of Hydrogen Generated by OTEC as Future Fuel. *International Journal of Hydrogen Energy*. 42 (1). pp. 26–37. https://doi.org/10.1016/j.ijhydene.2016.11.115.

(g)   World Bank. 2021. *Offshore Wind Roadmap for the Philippines*. Washington, DC.

(h)   N. Dinh. 2022. *Projections of Levelized Costs of Hydrogen (LCOH), MARES Report*. Manila: ADB.

(i)   Salinity gradient research and technologies in Palau need more exploration, investigation, and feasibility studies.

References

Amrouche, S. O. et al. 2016. Overview of Energy Storage in Renewable Energy Systems. *International Journal of Hydrogen Energy.* 41 (45). pp. 20914–20927.

Bax, N., et al. 2021. Ocean Resource Use: Building the Coastal Blue Economy. *Reviews in Fish Biology and Fisheries.* 32 (1). pp. 189–207.

Baykara, S. Z. 2018. Hydrogen: A Brief Overview of Its Sources, Production and Environmental Impact. *International Journal of Hydrogen Energy.* 43 (23). pp. 10605–10614.

Bertheau, P. et al. 2020. Assessment of Microgrid Potential in Southeast Asia Based on the Application of Geospatial and Microgrid Simulation and Planning Tools. In O. Gandhi and D. Srinivasan, eds. *Sustainable Energy Solutions for Remote Areas in the Tropics.* Switzerland: Springer. pp. 149–178.

Bramley, B. et al. 2021. *The Blue Economy in Practice – Raising Lives and Livelihoods.* Massachusetts: Blue Economy Pulse / NLA International.

———. 2010. *Conservation Tourism.* Wallingford: CAB International.

Buckley, R. C. 2009. *Ecotourism: Principles and Practices.* Wallingford: CAB International.

Cabico, G. K. 2018. SWS: 3.6M Filipino Families Experienced Hunger in Q4 2017. *Philstar Global.* 22 January.

Clément, A., et al. 2002. Wave Energy in Europe: Current Status and Perspectives. *Renewable and Sustainable Energy Reviews.* 6 (5). pp. 405–431.

Cowell, A. 2021. The Blue Economy – A Drop in the Ocean. *KPMG Blog.* 3 June.

Dao, T. 2018. Vietnam Poised to Become Top Player in Ocean Aquaculture. *Seafood Source.* 9 August.

De Beukelaer, C. 2020. Ships moved more than 11 billion tonnes of our stuff around the globe last year, and it's killing the climate. This week is a chance to change. *The Conversation.* 16 November.

Dincer, I. and C. Acar. 2015. Review and Evaluation of Hydrogen Production Methods for Better Sustainability. *International Journal of Hydrogen Energy.* 40 (34). pp. 11094–11111.

Dinh, V. N. and E. McKeogh. 2018. Offshore Wind Energy: Technology Opportunities and Challenges. In *Lecture Notes in Civil Engineering.* Vol. 18. Switzerland: Springer. pp. 3–22.

Doherty, B. 2019. Pacific Islands seek $500m to make ocean's shipping zero carbon. News release. *The Guardian.* 24 September.

Edwards, A. J., ed. 2010. *Reef Rehabilitation Manual.* St. Lucia, Australia: Coral Reef Targeted Research & Capacity Building for Management Program.

Energy Sector Management Assistance Program (ESMAP). 2019. *Going Global: Expanding Offshore Wind to Emerging Markets.* Washington, DC: World Bank.

European Commission. 2021. European Green Deal: Developing a Sustainable Blue Economy in the European Union. Press release. 17 May.

———. Our Oceans, Seas and Coasts.

———. Shipbuilding Sector.

Fennell, D. 2003. *Ecotourism.* London: Routledge.

Fletcher, R. 2018. Fish and Chips: How Computer Analytics Can Transform Aquaculture. *The Fish Site.* 17 September.

Food and Agriculture Organization of the United Nations (FAO). 2018. Is the planet approaching "peak fish"? Not so fast, study says. News release. 9 July.

———. 2020. *The State of World Fisheries and Aquaculture.* Rome.

———. 2021. *The State of World Fisheries and Aquaculture.* Rome.

Global Factor. 2018. *4th Interim Delivery Assessment: Assessing the Existence of Key Elements of Blue and Circular Economy in the Caribbean.* Bilbao, Spain: Factor. 347 pp.

Gokkon, B. 2018. Indonesian Fish Farmers Get Early-Warning System for Lake Pollution. *Mongabay Environmental News.* 25 September.

Government of Fiji. 2020. Decarbonising Domestic Shipping Industry: Pacific Blue Shipping Partnership. Suva: Ministry of Trade, Co-operatives, Small and Medium Enterprises.

Hein, M. Y. et al. 2020. *Coral Reef Restoration as a Strategy to Improve Ecosystem Services – A Guide to Coral Restoration Methods.* Nairobi: United Nations Environment Programme. 60 pp.

IHI Corporation. 2019. IHI Demonstrated the World's Largest Ocean Current Turbine for the First Time in the World. *IHI Engineering Review.* 52 (1). pp. 6–9.

International Energy Agency. 2019. *Offshore Wind Outlook 2019.* Paris.

International Renewable Energy Agency (IRENA). 2014. *Ocean Energy: Technology Readiness, Patents, Deployment Status and Outlook.* Abu Dhabi.

———. 2014. *Tidal Energy: Technology Brief*. Abu Dhabi.

———. 2016. *Innovation Outlook: Offshore Wind*. Abu Dhabi.

———. 2019. *Future of Wind: Deployment, Investment, Technology, Grid Integration and Socio-Economic Aspects*. Abu Dhabi.

———. 2020. *Innovation Outlook: Ocean Energy Technologies*. Abu Dhabi.

Jepma, C. J. 2015. *Smart Sustainable Combinations in the North Sea Area*. Groningen: Energy Delta Institute.

Jepma, C. J. and M. van Schot. 2017. On the Economics of Offshore Energy Conversion: Smart Combinations— Converting Offshore Wind Energy into Green Hydrogen on Existing Oil and Gas Platforms in the North Sea. Groningen: Energy Delta Institute.

Juneja, M. 2021. Blue Economy: An Ocean of Livelihood Opportunities in India. *The Energy and Resources Institute*. 12 March.

Karokaro, A. 2018. Another Mass Fish Kill Hits Indonesia's Largest Lake. *Mongabay Environmental News*. 24 August.

Kirezci, E. et al. 2020. Projections of Global-Scale Extreme Sea Levels and Resulting Episodic Coastal Flooding over the 21st Century. *Scientific Reports*. 10 (1). 11629.

KPMG. 2020. The Time Has Come: The KPMG Survey of Sustainability Reporting 2020.

Lam, V. W. Y. et al. 2020. Climate Change, Tropical Fisheries and Prospects for Sustainable Development. *Nature Reviews Earth & Environment*. 1 (9). pp. 440–454.

Lamy, Y. S., C. Leijonhufvud, and N. O'Donohoe. 2021. The Next 10 Years of Impact Investment. *Stanford Social Innovation Review*. 16 March.

Laurens, L. M. L. 2017. State of Technology Review – Algae Bioenergy. IEA Bioenergy Technology Collaboration Programme.

Mariani, G. et al. 2020. Let More Big Fish Sink: Fisheries Prevent Blue Carbon Sequestration—Half in Unprofitable Areas. *Science Advances*. 6 (44)

Martinez, M. et al. 2021. A Systemic View of Potential Environmental Impacts of Ocean Energy Production. *Renewable and Sustainable Energy Reviews*. 149. 111332.

Micronesian Center for Sustainable Transport. Pacific Blue Shipping Partnership.

Mora, D. and A. deRijck. 2015. Blue Energy: Salinity Gradient Power in Practice. *GSDR Brief*. Global Sustainable Development Report.

Newsome, D., S. A. Moore, and R. K. Dowling. 2002. *Natural Area Tourism: Ecology, Impacts and Management*. Clevedon: Channel View Publications.

O'Kelly-Lynch, P. D. et al. 2020. Offshore Conversion of Wind Power to Gaseous Fuels: Feasibility Study in a Depleted Gas Field. *The Journal of Power and Energy, Part A of the Proceedings of the Institution of Mechanical Engineers*. 234 (2). pp. 226–236.

Okafor-Yarwood, I. et al. 2020. The Blue Economy–Cultural Livelihood– Ecosystem Conservation Triangle: The African Experience. *Frontiers in Marine Science*. 7.

Organisation for Economic Co-operation and Development (OECD). 2016. *The Ocean Economy in 2030*. Paris: OECD Publishing.

———. Ocean Shipping and Shipbuilding.

Pacoureau, N. et al. 2021. Half a Century of Global Decline in Oceanic Sharks and Rays. *Nature*. 589 (7843). pp. 567–571.

Patil, P. G. et al. 2016. *Toward a Blue Economy: A Promise for Sustainable Growth in the Caribbean*. Washington, DC: World Bank.

Pauli, G. 2010. *The Blue Economy: 10 Years, 100 Innovations, 100 Million Jobs*. Berlin: Konvergenta Publishing UG.

Philippine Statistics Authority. 2019. Fisheries Situation Report for Major Species: January to December 2021.

Pushpoth, V., V. N. Dinh, and E. McKeogh. 2018. *Local Energy Storage for Offshore Windfarms*. School of Engineering, University College Cork. 20 April.

Raworth, K. 2017. *Doughnut Economics: Seven Ways to Think Like a 21st Century Economist*. London: Penguin Random House.

Reuß, M. et al. 2017. Seasonal Storage and Alternative Carriers: A Flexible Hydrogen Supply Chain Model. *Applied Energy*. 200. pp. 290–302.

Rogers, A. et al. 2020. *Critical Habitats and Biodiversity: Inventory, Thresholds and Governance*. Washington, DC: World Resources Institute.

Sharma, S. and S. K. Ghoshal. 2015. Hydrogen the Future Transportation Fuel: From Production to Applications. *Renewable and Sustainable Energy Reviews*. 43 (12). pp. 1151–1158.

Smith, H. and X. Basurto. 2019. Defining Small-Scale Fisheries and Examining the Role of Science in Shaping Perceptions of Who and What Counts: A Systematic Review. *Frontiers in Marine Science*. 6. 236.

Smith-Godfrey, S. 2016. Defining the Blue Economy. *Maritime Affairs: Journal of the National Maritime Foundation of India*. 12 (1). pp. 58–64.

Standing, G. 2022. *The Blue Commons: Rescuing the Economy of the Sea*. London: Pelican Books.

Steffen, W. et al. 2015. Planetary Boundaries: Guiding Human Development on a Changing Planet. *Science*. 347 (6223).

Sumaila, U. R. et al. 2021. Financing a Sustainable Ocean Economy. *Nature Communications*. 12 (1). 3259.

Takagi, T. 2019. Energy Production: Biomass – Marine. In M. Ueda, ed. *Yeast Cell Surface Engineering: Biological Mechanisms and Practical Applications*. Springer Singapore. pp. 29–41.

Tehthe, L. C. L. and U. R. Sumaila. 2011. Contribution of Marine Fisheries to Worldwide Employment. *Fish and Fisheries*. 14 (1).

The Economist Intelligence Unit. 2015. The Blue Economy: Growth, Opportunity and a Sustainable Ocean Economy. Briefing Paper for the World Ocean Summit 2015. Cascais, Portugal. 3–5 June.

US Department of Energy, Office of Energy Efficiency & Renewable Energy, Hydrogen and Fuel Cell Technologies Office. Hydrogen Storage.

US Department of Energy, Federal Energy Management Program. 2016. Ocean Energy. *Whole Building Design Guide*. 18 November.

US Soybean Export Council. 2018. Aquaculture is Fastest Growing Food Production Sector, According to FAO Report. 16 July.

United Nations Environment Programme (UNEP). 2012. *Green Economy in a Blue World*. Nairobi.

———. 2016. *Blue Economy Concept Paper*. Nairobi.

Weaver, D. 2008. *Ecotourism*. 2nd edition. Brisbane: John Wiley and Sons, Inc.

Witter, M. 2021. COVID-19: Intensifying the Existential Threat to the Caribbean. *Agrarian South: Journal of Political Economy*. 10 (1).

World Bank. 2017. *The Potential of the Blue Economy: Increasing Long-Term Benefits of the Sustainable Use of Marine Resources for Small Island Developing States and Coastal Least Developed Countries*. Washington, DC.

World Economic Forum. 2016. *The New Plastics Economy Rethinking the Future of Plastics*. Geneva.